Numbers: A Very Short Introduction

VERY SHORT INTRODUCTIONS are for anyone wanting a stimulating and accessible way in to a new subject. They are written by experts and have been translated into more than 40 different languages. The series began in 1995 and now covers a wide variety of topics in every discipline. The VSI library contains nearly 400 volumes—a Very Short Introduction to everything from Indian philosophy to psychology and American history—and continues to grow in every subject area.

Very Short Introductions available now:

Peter M. Higgins

NUMBERS

A Very Short Introduction

OXFORD
UNIVERSITY PRESS

OXFORD
UNIVERSITY PRESS

Great Clarendon Street, Oxford OX2 6DP

Oxford University Press is a department of the University of Oxford.
It furthers the University's objective of excellence in research, scholarship,
and education by publishing worldwide in

Oxford New York

Auckland Cape Town Dar es Salaam Hong Kong Karachi
Kuala Lumpur Madrid Melbourne Mexico City Nairobi
New Delhi Shanghai Taipei Toronto

With offices in

Argentina Austria Brazil Chile Czech Republic France Greece
Guatemala Hungary Italy Japan Poland Portugal Singapore
South Korea Switzerland Thailand Turkey Ukraine Vietnam

Oxford is a registered trade mark of Oxford University Press
in the UK and in certain other countries

Published in the United States
by Oxford University Press Inc., New York

British Library Cataloguing in Publication Data
Data available

Library of Congress Cataloging in Publication Data
Data available

Typeset by SPI Publisher Services, Pondicherry, India
Printed in Great Britain on acid-free paper by
Ashford Colour Press Ltd, Gosport, Hampshire

ISBN 978-0-19-958405-5

5 7 9 10 8 6

Contents

Preface

The purpose of this little book is to explain, in language that will be familiar to everyone, what are the various kinds of numbers that arise and how they behave. Numbers allow comparisons between all manner of things, and anyone with no understanding of numbers would be lost in the modern world where things numerical are there to greet us at every turn. We should realize however that, despite their familiarity, numbers have no physical existence but rather are abstractions that we elicit from the world we find ourselves in. To develop a clear picture of how they operate, it is sometimes better to consider them in their own right, without reference to anything else.

This *Short Introduction* is not a refresher course in arithmetic, nor will that much be said on the history of the number system. Rather its purpose is to explain numbers themselves and the kinds of behaviour they exhibit. A glance at the list of chapters reveals that the first part of the book deals mainly with ordinary counting numbers, while in the second half we go beyond that. By exploring natural problems that arise in commerce and science, the need to freely perform calculations eventually has taken us, by stages, into the arena of the complex numbers, which is the principal underlying framework for most number matters. This may sound a little daunting but rest assured, the hard work has already been done for you.

The modern number system did not come to us gift-wrapped but rather has developed over many centuries. There were long periods of confusion, which had two root causes. The first was the lack of an efficient way to represent the numbers we need that would allow us to manipulate them. The second, which was related to the first, was philosophical agonizing over the interpretations of various number types and whether or not they are meaningful. Nowadays we are much more sure of ourselves when it comes to what we do and don't need to worry about when dealing with numbers, making it possible to give a complete picture of the number world in a single short account like this one. That is not to say that all mystery has vanished – far from it, as you will discover as you read on.

Peter M. Higgins
Colchester, England, 2011

List of illustrations

Chapter 1
How not to think about numbers

We are all used to seeing numbers written down, and to drawing some meaning from them. However, a numeral such as 6 and the number that it represents are not one and the same thing. In Roman numerals, for example, we would write the number six as VI, but we realize that this stands for the same number that is written as 6 in modern notation. Both symbolize collections of the kind corresponding to six tally marks: I I I I I I. We shall first spend a little time considering the different ways that we represent and think about numbers.

We sometimes solve number problems almost without realizing it. For example, suppose you are conducting a meeting and you want to ensure that everyone there has a copy of the agenda. You can deal with this by labelling each copy of the handout in turn with the initials of each of those present. As long as you do not run out of copies before completing this process, you will know that you have a sufficient number to go around. You have then solved this problem without resorting to arithmetic and without explicit counting. There are numbers at work for us here all the same and they allow precise comparison of one collection with another, even though the members that make up the collections could have entirely different characters, as is the case here, where one set is a collection of people, while the other consists of pieces of paper.

What numbers allow us to do is to compare the relative size of one set with another.

In the previous scenario you need not bother to count how many people were present as you did not have to know – your problem was to determine whether or not the number of copies of the agenda was at least as great as the number of people, and the value of these numbers was not required. You will, however, need to take a count of the number present when you order lunch for fifteen and certainly, when it comes to totting up the bill for that meal, someone will make use of arithmetic to work out the exact cost, even if the sums are all done on a calculator.

Our modern number system allows us to express numbers in an efficient and uniform manner, which makes it easy to compare one number with another and to perform the arithmetical operations that arise through counting. In the day-to-day world, we employ base ten for all our arithmetic, that is to say we count by tens, and we do that for the accidental reason that we have ten digits on our hands. What makes our number system so effective, however, is not our particular choice of base but rather the use of *positional value* in number representation, where the value of a numeral depends on its place in the number string. For example, 1984 is short for 4 ones plus 8 lots of ten plus 9 hundreds plus 1 thousand.

It is important to understand what we mean when we write numbers in particular ways. In this chapter, we will think about what numbers represent, discover different approaches to counting, meet a very important set of numbers (the primes), and introduce some simple number tricks for finding them.

How counting was sorted out

It is worth taking a few moments to appreciate that there are two distinct stages to the process of building a counting system based on, for instance, tens. Two basic tasks that we impose on children are remembering how to recite the alphabet and learning how to count. These processes are superficially similar but yet have fundamental differences. Our language is based on a 26-letter alphabet and, roughly speaking, each letter corresponds to a sound that we use to speak words. In any case, it is certainly true that the English language has developed so that it can be written using a set of 26 symbols. However, we cannot compile dictionaries unless we assign an order to our alphabet. There is no particularly natural order available and the one that we have settled on and all sing in school, $a, b, c, d, \ldots$ seems very arbitrary indeed. To be sure, the more frequently used letters generally occur in the first half of the alphabet, but this is only a rough guide rather than a rule, with the common letters s and t, for example, sounding off late in the roll call. By contrast, the counting numbers, or *natural numbers* as they are called, $1, 2, 3, \ldots$ come to us in that order: for example, the symbol 3 is meant to stand for the number that follows 2 and so has to be listed as its successor. We can, up to a point, make up a fresh name for each successive number. Sooner or later, however, we have to give up and start grouping numbers in batches in order to handle the unending sequence. Grouping by tens represented the first stage of developing a sound number system, and this approach has been near universal throughout history and across the globe.

There was, however, much variation in detail. The Roman system favours gathering by fives as much as tens, with special symbols, V and L, for five and for fifty respectively. The Ancient Greek system was squarely based on grouping by tens. They would use specific letters to stand for numbers, sometimes dashed to tell the reader that the symbol should be read as a number rather than as a letter in some ordinary word. For example, π stood for 80 and γ for 3, so they might write $\pi\gamma$ to denote 83. This may look equally as efficient, and indeed much the same as our notation, but it is not. The Greeks still missed the point of the positional system as the value of each of their symbols was fixed. In particular, $\gamma\pi$ could still only represent the same number, 3 + 80, whereas if we switch the order of the digits in 83 we have the different number 38.

In the Hindu-Arabic system, the second stage of number representation was attained. Here the big idea is to make the value of a symbol dependent upon where it occurs in the string. This allows us to express any number with just a fixed family of symbols. We have settled on the set of ten numerals $0, 1, 2, \cdots, 9$, so the normal number system is described as *base ten*, but we could build our number system up from a larger or a smaller collection of basic symbols. We can even manage with as few as two numerals, 0 and 1 say, which is what is known as the *binary* system, so often used in computing. It is not the choice of base size, however, that was revolutionary but the idea of using position to convey extra information about the identity of your numbers.

For example, when we write a number like 1905, the value of each digit depends on its place in the number string. Here there are 5 units, 9 lots of one hundred (which is 10×10), and one lot of one thousand (which is $10 \times 10 \times 10$). The use of the zero symbol is important as a placeholder. In the case of 1905, no contribution comes from the 10's place, but we cannot ignore that and just write 195 instead, as that represents an entirely different number. Indeed, each string of digits represents a different number and it is for that reason that huge numbers may be represented by short strings. For instance, we can assign a unique number to every human being on Earth using strings no longer than ten digits and in this way give a personal identifier to every individual belonging to this huge set.

Societies of the past sometimes used different bases for their writing of numbers but that is much less significant than the fact that nearly all of them lacked a true positional system with full use of a zero symbol as a placeholder. In view of how very ancient is the civilization of Babylon, it is remarkable that they among the peoples of the ancient world came closest to a positional system.

They did not, however, fully embrace the use of the not-so-natural number 0 and eschewed using the empty register in the final position the way we do to distinguish, for example, 830 from 83.

The conceptual hurdle that had to be cleared was the realization that zero was indeed a number. Admittedly, zero is not a positive number but it is a number all the same and until we incorporate it into our number system in a fully consistent manner, we remain handicapped. This crucial final step was taken in India in about the 6th century AD. Our number system is called Hindu-Arabic as it was communicated from India to Europe via Arabia.

Living with and without decimals

Of course, we have now extended the base ten positional idea into fractional parts giving the familiar decimal number system. When we write 3.14, for instance, the 1 after the decimal point stands for one lot of $\frac{1}{10}$, a tenth part, and the 4 similarly stands for $\frac{4}{100}$. This two-decimal-place form of a number is very familiar to us as we deal in decimal currency where the smallest unit is not one dollar, or pound, or euro, as the case may be, but the 'penny' or 'cent', which is one-hundreth part of the main currency unit. Decimal arithmetic is the natural extension of the base ten system and in practice it represents the best way to carry out ordinary sums. Despite all its advantages, the decimal approach had a slow and hesitant genesis. It remained within the province of a mathematical elite until the latter part of the 16th century, when it finally found its way into commercial arithmetic and general usage. Even after that time, grouping based on numbers other than powers of ten persisted. Britain did not adopt decimal currency until 1971. And part of the English-speaking world still resolutely sticks with yards, feet, and inches. In defence of Imperial Units, they are convenient in size, being very much attuned to the human scale of things. Our hands are six to eight inches long and we are five to six feet tall, so we surround ourselves with objects of similar size, which are then readily measured in units of feet and inches. However, ten inches to the foot could have worked just as well and would have been more easily dealt with by our base ten calculators.

Adopting a particular base for a number system is a little like placing a particular grid scale on a map. It is not intrinsic to the object but is rather akin to a system of coordinates imposed on top as an instrument of control. Our choice of base is arbitrary in character and the exclusive use of base ten saddles us all with a blinkered view of the set of counting numbers: $1, 2, 3, 4, \cdots$. Only by lifting the veil can we see numbers face to face for what they truly are. When we mention a particular number, let us say for example forty-nine, all of us have a mental picture of the two numerals 49. This is somewhat unfair to the number in question as we are immediately typecasting forty-nine as $(4 \times 10) + 9$. Since $49 = (4 \times 12) + 1$, it may just as easily be thought of that way and, indeed, in base twelve, forty-nine would therefore be written as 41, with the numeral 4 now standing for 4 lots of 12. However, what gives the number forty-nine its character is that it equals the product 7×7, known as the *square* of 7. This facet of its personality is highlighted in base seven, as then the number forty-nine is represented as 100, the 1 now standing for one lot of 7×7.

We would be equally entitled to use another base, such as twelve, for our number system: the Mayans used twenty and the Babylonians base sixty. In one way, the number 60 is a good choice for a counting base as 60 has many divisors, being the smallest number divisible by all the numbers from 1 through to 6. A relatively large number such as 60 has the disadvantage, however, that to use it as a base would require us to introduce 60 separate symbols to stand for each of the numbers from zero up to fifty-nine.

One number is a *factor* of another if the first number divides into the second a whole number of times. For example, 6 is a factor of

$42 = 6 \times 7$ but 8 is not a factor of 28 as 8 into 28 goes 3 times with a remainder of 4. The property of having many factors is a handy one to have for the base of your number system, which is why twelve may have been a better choice than ten for our number base as 12 has 1, 2, 3, 4, 6, and 12 as its list of factors while 10 is divisible only by 1, 2, 5, and 10.

The effectiveness and sheer familiarity of our number system embues us with a false confidence and with some inhibitions. We feel happier with a single whole number than with an arithmetical expression. For example, most people would rather talk about 5969 than 47×127, although the two expressions represent the same thing. Only after 'working out the answer', 5969, do we feel that we 'have' the number and can look it in the eye. There is, however, an element of delusion in this as we have only written the number as a sum of powers of ten. The general shape of the number and other properties can be inferred more from the alternative form where the number is broken down into a product of factors. To be sure, this standard form, 5969, does allow direct comparison with other numbers that are expressed in the same way but it does not reveal the full nature of the number. In Chapter 4, you will see one reason why the factorized form of a number can be much more precious than its base ten representation, which can keep vital factors hidden.

One advantage that the ancients did have over us is that they were not mentally trapped within a decimal-style mindset. When it came to number patterns, it was natural for them to think in terms of special geometric properties that a particular number may or may not enjoy. For example, numbers such as 10 and 15 are *triangular*, something that is visible to us through the triangle of pins in ten-pin bowling and the triangular rack of fifteen red balls in snooker. But this is not something that comes to mind from the base ten displays of these numbers alone. The freedom the ancients enjoyed by default we can recapture by casting aside our base ten prejudices and telling ourselves that we are free to think of numbers in quite different ways.

Having emancipated ourselves in this way, we might choose to focus on factorizations of a number, that is to say the way the number can be written as a product of smaller numbers multiplied together. Factorizations reveal something of the number's inner structure. If we suspend the habit of thinking of numbers simply as servants of science and commerce, and take a little time to study them in their own right without reference to anything else, much is revealed that otherwise would remain hidden. The natures of individual numbers can manifest themselves in ordered patterns in nature, more subtle than mere triangles and squares, like the spiral head of a sunflower, which represents a so-called Fibonacci number, a number type that will be introduced in Chapter 5.

A glance at the prime number sequence

One of the glories of numbers is so self-evident that it may easily be overlooked – every one of them is unique. Each number has its own structure, its own character if you like, and the personality of individual numbers is important because when a particular number arises, its nature has consequences for the structure of the collection to which that number applies. There are also relationships between numbers that reveal themselves when we carry out the fundamental number operations of addition and multiplication. Clearly any counting number greater than 1 can be expressed as the sum of smaller numbers. However, when we start multiplying numbers together, we soon notice that there are some numbers that never turn up as the answer to our sums. These numbers are the *primes* and they represent the building blocks of multiplication.

A *prime number* is a number like 7 or 23 or 103, which has exactly two factors, those necessarily being 1 and the number itself. (The word *divisor* is also used as an alternative word for factor.) We do not count 1 as a prime as it has only one factor. The first prime then is 2, which is the only even prime and the following trio of odd numbers 3, 5, and 7 are all prime. Numbers greater than 1 that are not prime are called *composite* as they are composed of smaller numbers. The number $4 = 2 \times 2 = 2^2$ is the first composite number; 9 is the first odd composite number, and $9 = 3^2$ is also a square. With the number $6 = 2 \times 3$, we have the first truly composite number in that it is composed of two different factors that are greater than 1 but smaller than the number itself, while $8 = 2^3$ is the first proper *cube*, which is the word that means that the number is equal to some number raised to the power 3.

After the single-digit numbers, we have our chosen number base $10 = 2 \times 5$, which is special nonetheless being *triangular* in that $10 = 1 + 2 + 3 + 4$ (remember ten-pin bowling). We then have a pair of *twin primes* in 11 and 13, which are two consecutive odd numbers that are both prime, separated by the number 12, which in contrast has many factors for its size. Indeed, 12 is the first so-called *abundant number*, as the the sum of its *proper factors*, those less than the number itself, exceeds the number in question: $1 + 2 + 3 + 4 + 6 = 16$. The number $14 = 2 \times 7$ may look undistinguished but, as the paradoxical quip goes, being the *first* undistinguished number makes it distinguished after all. In $15 = 3 \times 5$, we have another triangular number and it is the first odd number that is the product of two proper factors. Of course, $16 = 2^4$ is not only a square but the first fourth power (after 1), making it very special indeed. The pair 17 and 19 are another pair of twin primes, and I leave the reader to make their own observations about the peculiar nature of the numbers 18, 20, and so on. For each you can make a claim to fame.

Returning to the primes, the first twenty of them are:

2, 3, 5, 7, 11, 13, 17, 19, 23, 29, 31, 37, 41, 43, 47, 53, 59, 61, 67, 71.

Clearly, near the very beginning of the number sequence, primes are commonplace as there is little opportunity for small numbers to have factorizations. After that, the primes become rarer. For example, there is only one triple of consecutive odd primes: the trio 3, 5, 7 is unique, as every third odd number is a multiple of 3, and so this can never happen again. The thinning process of prime occurrence is, however, quite leisurely and surprisingly erratic. For example, the thirties have only two primes, those being 31 and 37, yet immediately after 100 there are two 'consecutive' pairs of twin primes in 101, 103 and 107, 109.

The primes have been a source of fascination for thousands of years because they never run out (a claim that we shall justify in the next chapter) yet they arise among the natural numbers in a somewhat haphazard fashion. This mysterious and unpredictable facet of their nature is exploited in modern cryptography to safeguard confidential communication on the Internet, which is the subject of Chapter 4.

Checking for primality: prime divisibility tests

The most simple-minded way of finding all the primes up to a given number such as 100 is to write all the numbers down and cross off the composite numbers as you find them. The standard method based on this idea is called the *Sieve of Eratosthenes* and runs as follows. Begin by circling 2 and then cross off all the multiples of 2 (the other even numbers) in your list. Then return to the beginning, circle the first number you meet that has not been crossed off (which will be 3) and then cross off all its multiples in the remaining list. By repeating this process sufficiently often, the primes will emerge as those numbers that never get crossed out, although some will be circled and some not. For example, Figure 1 shows the workings of the sieve up to 60.

How do you know when you can stop sieving? You need to repeat this process until you circle a number that is greater than the square root of the largest number in your list. For instance, if you do your own sieve for all numbers up to 120, you will need to run

1. Prime sieve: the primes up to 60 are the numbers not crossed out

through the sieve for multiples of 2, 3, 5, and 7, and when you circle 11 you can stop, as $11^2 = 121$. At that point, you will have circled as far as the first prime exceeding the square root of your biggest number (120 in this case) with the remaining primes sitting there untouched. All composite numbers will now have been crossed out as each is a multiple of one or more of 2, 3, 5, and 7.

It is not hard to see why the square root of the greatest number n in your list determines the number of passes you need to make. (When explaining properties of arbitrary numbers, mathematicians give names, in the form of symbols, for the subjects of the discussion. For numbers, these names are typically lower-case letters, such as m and n; the product of two numbers $m \times n$ is often abbreviated as mn.) Any listed composite number m will have a prime factor and its *smallest* prime factor must be no more than the square root of n, because the product of two or more numbers that exceed $\sqrt{n}$ is greater than n (and so also greater than m).

Another aspect of the question of primality is whether a particular given number n is prime or composite. To decide that, we can test n for division by each of the primes in turn up to $\sqrt{n}$ and if n passes all these tests it will be prime, and otherwise not. For that reason, it is handy to have some simple ways of testing for divisibility by each of the small primes, 2, 3, 5, 7, ... This need is easily catered for as follows.

11

It is very easy to test for divisibility by 2 and by 5 as these primes are the prime factors of our number base ten. In view of this, you only need to check the final digit of the number n in question: n is divisible by 2 exactly when its units digit is even (i.e. 0, 2, 4, 6, or 8), and n has 5 as a factor if and only if it ends in 0 or 5. No matter how many digits the number n has, we only need to check the last digit to determine whether we have a multiple of 2 or of 5. For primes that do not divide into 10, we need to do a bit more work but nevertheless there are simple tests for divisibility that are much quicker than resorting to doing the full division sum.

A number is divisible by 3 if and only if the same is true of the sum of its digits. For example, the sum of the digits of $n = 145, 373, 270, 099, 876, 790$ is 87 and $87 = 3 \times 29$ and so n is in this case divisible by 3. Of course, we can apply the test to the number 87 itself and indeed go on taking the sum of digits of the outcome at each stage until the result is obvious. Doing this for the given example produces the following sequence:

$$145, 373, 270, 099, 876, 790 \rightarrow 87 \rightarrow 15 \rightarrow 6 = 2 \times 3.$$

You will see that all the division tests listed here are so quick that you can handle numbers with dozens of digits with relative ease even though these numbers are billions of times greater than the biggest number with which your hand calculator can cope.

The tests given here for the remaining primes up to 20 are chosen because they are all of the same general type. These routines are all simple to apply, although it is less obvious why they work. Although the justifications are not recorded here, the proofs of their validity are not especially difficult.

Let's begin with a test for divisibility of a given number n by 7. Double the final digit of n and subtract that from the number that remains when that final digit is removed. The new number will be a multiple of 7 precisely when the same is true of n. We repeat this process until the outcome is obvious. As a simple example, let us take $n = 3465$: twice 5 is 10 so we take 10 from the 346 to get 336; next we go again, taking twice 6, which is 12 from the 33 to get $21 = 3 \times 7$, and so n is divisible by 7. If you have forgotten your seven times tables, we can go through again: subtracting twice 1 from 2 leaves us with 0, which is divisible by 7, as $\frac{0}{7} = 0$. (It is fine to divide zero by a whole number – it is the reverse, dividing by zero, that has no meaning.) Even a number in the tens of millions can be dealt with easily in this way. In this and subsequent examples, we simply list the output number at each stage of the *algorithm*, which is the name given for a mechanical process such as this one that solves a given type of problem.

$$n = 27,916,924 \rightarrow 2,791,684 \rightarrow 279,160 \rightarrow 27,916 \rightarrow$$
$$2,779 \rightarrow 259 \rightarrow 7$$

and so n is divisible by 7. Each time we run through the loop of instructions, we lose at least one digit, so the number of passes through the loop is about the same as the length of the number with which we begin.

To test whether or not n has a factor of 11, subtract the final digit from the remaining truncated number and repeat. For example, the next number is a multiple of 11 as our method reveals:

$$4,959,746 \rightarrow 495,968 \rightarrow 49,588 \rightarrow 4,950 \rightarrow 495$$
$$\rightarrow 44 = 4 \times 11.$$

To check for divisibility by 13, *add* four times the final digit to the remaining truncated number and precede as with 7 and 11. For instance, the next number turns out to have 13 as one of its prime factors:

$$11,264,331 \rightarrow 1,126,437 \rightarrow 112,671 \rightarrow 11,271$$
$$\rightarrow 1131 \rightarrow 117 \rightarrow 39 = 3 \times 13.$$

For 17 and for 19, we subtract five times the final digit in the case of 17, and add twice the final digit when testing if 19 is a factor, once more applying this step to the truncated number that

13

remains, repeating the process as often as we need. For example, we test 18, 905 for divisibility by 17:

$$18, 905 \rightarrow 1, 865 \rightarrow 161 \rightarrow 11$$

so it is not a multiple of 17, but for 19, the test gives the opposite conclusion:

$$18, 905 \rightarrow 1, 900 = 100 \times 19.$$

Armed with this battery of tests, you can readily check the primality of all numbers up to 500 (as $23^2 = 529$ exceeds 500, so 19 is the largest potential prime factor that you need concern yourself with). For example, to settle the matter for 247, we just need to check for divisibility up to the prime 13 (as the square of the next prime, $17^2 = 289$, exceeds 247). Applying the test for 13, however, we learn from $247 \rightarrow (24 + 28) = 52 \rightarrow 13$, that we have a multiple of 13: ($247 = 19 \times 13$).

The divisibility tests for primes can also be mounted in parallel to furnish divisibility tests for those numbers that are *square-free* products of these primes (numbers not divisible by the square of any prime) such as $42 = 2 \times 3 \times 7$: a number n will be divisible by 42 exactly when n passes the trio of divisibility tests for 2, 3, and 7. However, tests for those numbers that have square factors, such as $9 = 3^2$, do not come automatically, although it is the case that n has 9 as a factor if and only if that is true of the sum of the digits of n.

You might ask, after thousands of years, haven't those clever mathematicians come up with better and more sophisticated methods of testing for primality? The answer is yes. In 2002, a relatively quick way was discovered to test if a given number is prime. The so-called 'AKS primality test' does not, however, provide the factorization of the given number if it happens to be composite. The problem of finding the prime factors of a given number, although in principle solvable by trial, still seems practically intractable for extremely large integers, and for that reason it forms the basis of much ordinary encryption on the Internet, a subject to which we will return in Chapter 4. Before that we shall, in the next two chapters, look a little more closely at primes and factorization.

Chapter 2
The unending sequence
of primes

How primes fit into the number jigsaw

How can we be sure that the primes do not become rarer and rarer and eventually peter out altogether? You might think that since there are infinitely many counting numbers and each can be broken down into a product of primes (something explained more carefully in a moment), there must then be infinitely many primes to do the job. Although this conclusion is true, it does not follow from the previous observations, for if we begin with a finite collection of primes, there is no end to the number of different numbers we can produce just using those given prime factors. Indeed, there are infinitely many different powers of any single prime: for example, the powers of the prime 2 are 2, 4, 8, 16, 32, 64, $\cdots$. It is conceivable therefore that there are only finitely many primes and *every* number is a product of powers of those primes. What is more, we have no way of producing an unending series of different primes the way we can, for example, produce any number of squares, or multiples of a specific given number. When it comes to primes, we still have to go out hunting for them, so how can we be sure they do not become extinct?

We will all be sure by the end of this chapter, but first I will draw your attention to one simple 'pattern' among the primes worth noting. Every prime number, apart from 2 and 3, lies one side or the other of a multiple of 6. In other words, any prime after these first two has the form $6n \pm 1$ for some number n. (Remember that $6n$ is short for $6 \times n$ and the double symbol $\pm$ means plus or minus.) The reason for this is readily explained. Every number can be written in exactly one of the six forms $6n$, $6n \pm 1$, $6n \pm 2$, or $6n + 3$ as no number is more than three places away from some multiple of six. For example, $17 = (6 \times 3) - 1$, $28 = (6 \times 5) - 2$, $57 = (6 \times 9) + 3$; indeed, the six given forms appear in cyclic order, meaning that if you write down *any* six consecutive numbers, each of the forms will appear exactly once, after which they will reappear again and again, in the same order. It is evident that numbers of the forms $6n$ and $6n \pm 2$ are even, while any number of the form $6n + 3$ is divisible by 3. Therefore, with the obvious exceptions of 2 and 3, only numbers of the form $6n \pm 1$ can be prime. The case where *both* of the numbers $6n \pm 1$ are prime corresponds exactly to the twin primes: for example $(6 \times 18) \pm 1$ gives the pair 107, 109 mentioned in the first chapter. You might be tempted to conjecture that *at least one of* the two numbers $6n \pm 1$ is always prime – this is certainly true for the list of primes up to 100 but the first failure is not far away: $(6 \times 20) - 1 = 119 = 7 \times 19$, while $(6 \times 20) + 1 = 121 = 11^2$, so neither number is prime when we take $n = 20$.

And the principal reason why primes are important is that every number can be written as a product of prime numbers, and that can be done in essentially only one way. To find this special factorization, we just need to factorize the given number in some way and then continue factorizing any composite factors that appear until this can be done no more. For example, we could say that $120 = 2 \times 60$ and continue by breaking the composite factor of 60 down further to give:

$$120 = 2 \times 60 = 2 \times (2 \times 30) = 2 \times 2 \times (2 \times 15) = 2 \times 2 \times 2 \times 3 \times 5.$$

We say that the *prime factorization* of 120 is $2^3 \times 3 \times 5$. We could, however, have came to this by another route. For instance

$$120 = 12 \times 10 = (3 \times 4) \times (2 \times 5) = (3 \times (2 \times 2)) \times (2 \times 5)$$

but rearranging the prime factors from least to greatest still yields the same result as before: $120 = 2^3 \times 3 \times 5$.

At least it did in that example, and this behaviour may be more or less familiar to you, but how can you be sure that this applies to every number? It is clear enough that any number can be broken down into a product of primes but, since there is in general more than one way of tackling this task, how can we be sure that the process will always deliver the same final result? This is an important question, so I will take a few moments to give an outline of the reasoning that allows us to be absolutely sure about this. It is a consequence of another special property of prime numbers that we shall call the *euclidean property*: if a prime number is a factor of a product of two or more numbers, then it is a factor of one of the numbers in that product. For example, 7 is a factor of $8 \times 35 = 280$ (as the product $280 = 7 \times 40$) and we note that 7 is a factor of 35. This property characterizes primes as no composite number can give you the same guarantee: for example, we see that 6 is a factor of $8 \times 15 = 120$ (as $120 = 6 \times 20$) yet 6 is not a factor of either 8 or 15.

The fact that primes always have the above property can be proved using an argument based on what is known as the Euclidean Algorithm, which will be explained in Chapter 4. If we take this on trust for the time being, it is not too difficult to explain why no number could have two different prime factorizations, for suppose there were such a number. There then would be a smallest one that behaved in this way: let us denote it by n and so n has two prime factorizations which, when the prime factors are written in ascending order, are *not* identical. We shall show that this leads to contradiction and so must be false.

If these two factorizations of n had a prime p in common, we could cancel p from both and obtain two different prime factorizations of the smaller number $\frac{n}{p}$. Since n is the smallest number with two distinct prime factorizations, this is not possible and so the sets of primes involved in each factorization of n must have no prime in common. Now take one prime p that occurs in one of these factorizations of n. Since this prime p is a factor of n, it follows that p is a factor of the second factorization, and so, by the euclidean property, p is a factor of one of the primes, q say, of the second factorization. *But since q and p are both prime*, this is only possible if $q = p$, a possibility that we have already discounted as the two factorizations of n have no common prime. And so we arrive at a final contradiction, showing that it is impossible for such a number n to exist. Therefore we conclude that the prime factorization of every number is unique.

It is worth noting that the uniqueness of prime factorization would not hold if we included the number 1 among the primes, as we can adjoin any power of 1 to a factorization and the product retains the same value. This shows that 1 is fundamentally different in nature to the primes, and so it is right to frame the definition of prime number in a way that excludes 1 from the collection.

Euclid's infinity of primes

Let us return to the question as to how we know that the primes go on forever and that there is no way past them. If someone claimed that 101 is the largest prime, you can refute him at once by showing that 103 has no factors (except for 1 and 103) and so 103 is a larger prime. Your friend might then concede that he made a slip and that he should have said that it was 103 that is the largest prime of all. You could then show him up again by demonstrating that 107 is also prime, but your friend might still persist in his error by adjusting his position to the largest prime number on view. He could even retreat a little further and admit that he does not know the identity of the largest prime but nevertheless continue to claim that he is certain that there is one.

The best way to settle this question would be to show that, given any conceivable finite collection of primes, we can produce a new prime not in the list. For example, if someone claimed there was a largest odd number out there somewhere, you could refute him by saying that if n is odd, then $n + 2$ is a larger odd number, so there cannot be a largest odd number. This approach, however, is not so easy for the primes – given a finite list of primes, we have no way of using the collection to manufacture a prime that is demonstrably bigger than all of them. Perhaps there is a biggest prime after all? How are we to know that our stubborn friend isn't right?

Euclid of Alexandria (*c*. 300 BC), the Greek mathematician and father of all things euclidean, did however know. Given a list $p_1, p_2, \cdots, p_k$ where each of the p_i denotes a different prime, he could not find a way of generating a new prime, so he reverted to an argument that is one step more subtle. He showed that there must be one or more new primes *within a certain range of numbers* (but his argument does not allow us to locate exactly where to find a prime within that range).

It goes like this. Let $p_1, p_2, \cdots, p_k$ be the list of the first k primes say, and consider the number n that is one more than the product of all these primes, so that $n = p_1 p_2 \cdots p_k + 1$. Either n is a prime, or is divisible by a prime smaller than itself, which cannot be any of $p_1, p_2, \cdots, p_k$, as if p is any one of these primes, then dividing n by p will leave a remainder of 1. It follows that any prime divisor of n is a new prime that is greater that all the primes $p_1, p_2, \cdots, p_k$ and no more than n itself. In particular, it follows from this that there can be no finite list of primes that contains every prime number, and so the sequence of primes continues on forever and will never be exhausted. Euclid's eternal proof of the infinity of primes is among the most admired in all of mathematics.

Although Euclid's argument does not tell exactly where to find the next prime number, the overall frequency of the primes is now quite well understood. For example, if we take any two numbers, a and b say, with no common factor and consider the sequence $a, a + b, a + 2b, a + 3b, \cdots$, it was shown by the German mathematician Johann Dirichlet (1805–59) that infinitely many members of such a sequence are prime. (Of course, there is no hope if a and b do have a common factor, d say, as then every member in the list is also a multiple of d, and so is not prime.) When $a = 1$ and $b = 2$, we get the sequence of odd numbers which we know, by Euclid's proof, contains infinitely many prime numbers. Indeed, it can be shown through fairly simple adaptations of Euclid's argument that other special cases such as the sequence of numbers of the forms $3 + 4n$, $5 + 6n$, and $5 + 8n$ (as n runs through the successive values $1, 2, 3, \cdots$), each have infinitely many primes. The general result of Dirichlet is, however, very difficult to prove.

Another simply stated result is that there is always at least one prime number greater than any given number n but less than $2n$ (for $n \geq 2$). (As an aid to memory, inequality signs such as this one, which stands for greater than or equal to, always point to the smaller quantity.) This fact, historically known as *Bertrand's Postulate*, can be proved using quite elementary mathematics, although the proof is itself quite tricky. We can verify the postulate for n up to 4000 by making use of the following list of primes. First observe that each number in the list after the initial prime 2 is smaller than twice its predecessor:

$$2, 3, 5, 7, 13, 23, 43, 83, 163, 317, 631, 1259, 2503, 4001.$$

For each n in the range up to 4000, take the largest prime p in the list that is no more than n; the next prime q then lies in the range $n < q < 2n$ and this then ensures that Bertrand's Postulate holds for all n up to 4000. For example, for $n = 100$, $p = 83$, and then $q = 163 < 2 \times 100$. A subtle argument involving the size of the so-called central binomial coefficients (introduced in Chapter 5) then shows that the postulate is also true for n larger than 4000.

However, we do not have to go too far before meeting similar-sounding problems that as yet remain unsolved. For example, no one knows if there is always a prime between any two consecutive squares. Another observation is that there seems to be enough primes to ensure that every even number n greater than 2 is the sum of two of them (*Goldbach's Conjecture*). This has been directly verified for n up to 10^{18}. We might then hope for a proof along the lines above for Bertrand's Postulate, where we show that beyond a certain specified integer N, we may introduce a comparison based on what is known about the distribution of the primes to ensure that there will always be at least one solution in primes p, q to the equation $p + q = 2n$ for any even number $2n \geq N$. This still eludes us, although there are weaker results along these lines – for example, it has been known since 1939 that every sufficiently large odd number is the sum of at most three prime numbers and that every even number is the sum of no more than 300,000 primes. Proof of the full Goldbach Conjecture still seems a long way off.

A simple result that has something of the flavour of the argument type referred to above is that there is a number n less than 4 billion that can be written as the sum of four different cubes in ten distinct ways. It is known that $1729 = 1^3 + 12^3 = 9^3 + 10^3$ is the smallest number that is the sum of *two* cubes in two different ways. However, we do not necessarily have to identify the number n in order to know that it must exist. Sometimes it is possible to know for certain that there are solutions to a problem, without actually finding any of those solutions explicitly.

In this case, we begin by noting that if we take four different numbers no more than a fixed integer m and form the sum of their cubes, the result is less than $4m^3$. However, if $m = 1000$, then an elementary calculation shows that the number of sums of four different cubes is more than 10 times the number $4m^3$, from which it follows that *some* number $n \leq 4m^3 = 4,000,000,000$ must be the sum of four cubes in at least ten different ways. The details involve calculations using binomial coefficients (introduced in Chapter 5) and are not especially difficult.

The global picture of prime distribution is summed up by the observation of the leading 19th-century German mathematician and physicist Karl Friedrich Gauss (1777–1855) that $p(n)$, the number of primes up to the number n, is approximately given by $n/\log n$ and that the approximation becomes more and more accurate as n increases. For example, if we take n to be one million, the ratio of $n/\log n$ suggests that, up to that stage, about one number in every 12.7 should be prime. Gauss's observation, which in detail says something more precise, was not proved until 1896. The logarithm function referred to here is the so-called *natural logarithm*, which is not based on powers of 10, but rather on powers of a special number e, which is approximately equal to 2.718. We shall hear more of this very famous number e in Chapter 6.

The most celebrated undecided question in number theory is the Riemann Hypothesis, which can only be explained in terms of complex numbers, which we have yet to introduce. However, I mention it here as the object of the question can be reformulated using the uniqueness of prime factorization to involve a certain infinite product featuring all the primes. This leads to an interpretation which says that the Hypothesis implies that the overall distribution of the primes is very regular in that, in the long run, primality will apparently occur randomly. Of course, whether or not a particular number is a prime is not a random event but what is meant is that primality, in the realm of the very large, takes on the mantle of randomness, with no additional pattern or structure to emerge. Many a number theorist has a heartfelt wish to see this 150-year-old conjecture settled in their own lifetime.

Since they represent so natural a sequence, it is almost irresistible to search for patterns among the primes. There are, however, no genuinely useful formulas for prime numbers. That is to say, there is no known rule that allows you to generate all prime numbers or even to calculate a sequence that consists entirely of different primes. There are some neat formulas but they are of little practical worth, some of them even require knowledge of the prime sequence to calculate their value so that they are essentially a cheat. Expressions such as $n^2 + n + 41$ are known as *polynomials*, and this one is a particularly rich source of primes. For example, putting $n = 1, 7$, and 20 in turn yields the primes 43, 107, and 461 respectively. Indeed, the output of this expression is prime for all values of n from $n = 0$ to $n = 39$. At the same time, however, it is clear that this polynomial will let us down when we put $n = 41$, as the result will have 41 as a factor, and indeed it fails for $n = 40$ as

$$40^2 + 40 + 41 = 40(40 + 1) + 41 = 40 \times 41 + 41 = (40 + 1)41 = 41^2.$$

In general, it is quite straightforward to show that no polynomial of this kind can yield a formula for primes, even if we allow powers higher than 2 to enter the expression.

It is possible to devise tests for primality of a number that can be stated in a few words. However, to be of use they would need to be quicker, at least in some cases, than the direct verification procedure described in Chapter 1. A famous result goes by the name of Wilson's Theorem. Its statement involves the use of numbers called *factorials*, which we will meet again in Chapter 5. The number $n!$, read 'n factorial', is just the product of all numbers up to n. For example, $5! = 5 \times 4 \times 3 \times 2 = 120$. Wilson's Theorem is then a very succinct statement: a number p is prime if and only if p is a factor of $1 + (p - 1)!$.

The proof of this result is not very difficult, and indeed in one direction it is nearly obvious: if p were composite, so that $p = ab$ say, then since both a and b are less than p, they each occur as factors of $(p - 1)!$ and so p is a divisor of this factorial as well. It follows that when we divide $1 + (p - 1)!$ by p, we will obtain a remainder of 1. (The case where $a = b$ requires a little more thought.) This is very reminiscent of Euclid's proof for the infinity of primes. It follows that *if p is a factor of* $1 + (p - 1)!$ *then p* must be prime. The converse is a little harder to prove: *if p is prime then p* is a factor of $1 + (p - 1)!$. This, however, is the surprising direction of the theorem, although the reader can easily verify particular cases: for example, the prime 5 is indeed a factor of $1 + 4! = 1 + 24 = 25$.

What Wilson's Theorem does is convert the problem of determining whether or not p is prime from a series of division problems (checking division by all primes up to $\sqrt{p}$) into a single division problem. However, the subject of the division, $1 + (p - 1)!$, is huge even for quite small values of p. Despite being a concise statement, Wilson's Theorem is of no real use in identifying particular prime numbers. For example, to check that 13 is prime by Wilson would require us to verify that 13 is a factor of $1 + 12! = 479, 001, 601$. (But by applying the divisibility test for 13 in Chapter 1, the reader can check that Wilson was right!) Compare this to the labour involved in simply checking that 13 is divisible by neither 2 nor 3. Although Wilson's Theorem is not useful in prime verification, it has more than ornamental value and can be used to demonstrate other theoretical results.

As a final observation, we can exploit factorials, which by design have many factors, in order to prove that no *arithmetic* sequence of numbers, that is to say one of the form $a, a + b, a + 2b, a + 3b, \cdots$ can consist *only* of primes as it is possible to show that the gap between successive primes can be arbitrarily large while the common difference between consecutive members of the previous sequence is fixed at b. To see this, consider the sequence of n consecutive integers:

$$(n + 1)! + 2, \ (n + 1)! + 3, \ (n + 1)! + 4, \cdots, (n + 1)! + n + 1.$$

Each of these numbers is composite, as the first is divisible by 2 (as each of the terms has 2 as a factor), the second is divisible by 3, the next by 4, and so on up until the final one in the list, which has

$n + 1$ as a factor. We therefore have, for any given n, a sequence of n consecutive numbers, none of which are prime.

Instead of focusing on numbers with the fewest possible factors (the primes), we shall in the next chapter turn to numbers with many factors, although we shall discover that here too there are surprising links to some very special prime numbers.

Chapter 3
Perfect and not so perfect numbers

Perfection in a number

It is often easy to find peculiar properties of small numbers that characterize them – for instance, 3 is the only number that is the sum of all the previous numbers, while 2 is the only even prime (making it the oddest prime of all). The number 6 has a truly unique property in that it is both the sum and product of all of its smaller factors: $6 = 1 + 2 + 3 = 1 \times 2 \times 3$.

The Pythagoreans called a number like 6 *perfect*, meaning that the number is the sum of its proper factors, as we shall call them, which are the divisors strictly smaller than the number itself. This kind of perfection is indeed very rare. The first five perfect numbers are 6, 28, 496, 8128, and 33,550,336. A lot is known about the even perfect numbers but, to this day, no one has been able to answer the basic question of the Ancients as to whether there are infinitely many of these special numbers. What is more, no one has found an odd one, nor proved that there are none. Any odd perfect number must be extremely large and there is a long list of special properties that such a number must possess in consequence of its odd perfection. However, all these restrictions have not as yet legislated such a number out of existence – conceivably, these special properties serve to direct our search for the elusive first odd perfect number, which may yet be awaiting discovery.

The even perfects were known to Euclid to have a tight connection with a very special sequence of primes, known to us as the *Mersenne primes* named after Marin Mersenne (1588–1648), a 17th-century French monk.

A *Mersenne number m* is one of the form $2^p - 1$, where p is itself a prime. If you take, by way of example, the first four primes, 2, 3, 5, and 7, the first four Mersenne numbers are seen to be: 3, 7, 31, and 127, which the reader can quickly verify as prime. If p were not prime, suppose $p = ab$ say, then $m = 2^p - 1$ is certainly not prime either, as it can be verified that in these circumstances the number m has $2^a - 1$ as a factor. However, if p is prime then the corresponding Mersenne number is often a prime, or so it seems.

And Euclid explained, back in 300 BC, that once you have a prime Mersenne number then there is a perfect number that goes with it, that number being $P = 2^{p-1}(2^p - 1)$. The reader can soon verify that the first four Mersenne primes do indeed give the first four perfect numbers listed above: for example, using the third prime 5 as our seed we get the perfect number $P = 2^4(2^5 - 1) = 16 \times 31 = 496$, the third perfect number in the previous list. (The factors of P are the powers of 2 up to 2^{p-1}, together with the same list of numbers multiplied by the prime $2^p - 1$. It is now an exercise in summing what are known as geometric series (explained in Chapter 5) to check that the proper factors of P do indeed sum to P.)

What is more, in the 18th century the great Swiss mathematician Leonhard Euler (1707–83) (pronounced 'Oiler') proved the reverse implication in that every even perfect number is of this type. In this way, Euclid and Euler together established a one-to-one match between the Mersenne primes and the even perfect numbers. However, the next natural question is, are all the Mersenne numbers prime? Sadly not, and failure is close at hand as the fifth Mersenne number equals $2^{11} - 1 = 2,047 = 23 \times 89$. Indeed, we do not even know if the sequence of Mersenne primes runs out or not – perhaps after a certain point all the Mersenne numbers will turn out to be composite.

The Mersenne numbers are natural prime candidates all the same, as it can be shown that any proper divisor, if one exists, of a Mersenne number m has the very special form $2kp + 1$. For example, when $p = 11$, by dent of this result, we need only check for division by primes of the form $22k + 1$. The two prime factors, 23 and 89, correspond to the values $k = 1$ and $k = 4$ respectively. This fact about divisors of Mersenne numbers also provides a bonus in that it affords us a second way of seeing that there must be infinitely many primes, for it shows that the smallest prime divisor of $2^p - 1$ exceeds p, and so p cannot be the largest prime. Since this applies to every prime p, we conclude that there is no largest prime and the prime sequence runs on forever.

Since we have no way of producing primes at will, there is, at any one time, a largest known prime and nowadays the champion is always a Mersenne prime, thanks to the international GIMPS venture (Great Internet Mersenne Prime Search). This is a collaborative project of volunteers, which began in 1996. The project uses thousands of personal computers working in parallel, which test Mersenne numbers for primality using a specially devised cocktail of tailor-made algorithms. The current world champion, announced in August 2008, is $2^p - 1$ where $p = 43, 112, 609$, although a new Mersenne prime was found in April 2009 with $p = 42, 643, 801$. These numbers have about 13 million digits and would take thousands of pages to write down in ordinary base ten notation.

Less than perfect numbers

Traditional number lore often focused on individual numbers thought to have special, if not magical, properties such as those that are perfect. However, a number *pair* with a similar trait is 220 and 284, the first *amicable pair*, meaning that the proper factors of each sums to the other – a kind of perfection extended to a couple. The renowned amateur French mathematician Pierre de Fermat (1601–65) found other amicable pairs, such as 17,296 and 18,416, while Euler discovered dozens more. Surprisingly, they both missed the small pair of 1184 and 1210, found by 16-year-old Nicolò Paganini in 1866. We can of course try to go beyond pairs and look for perfect triples, quadruples, and so on. Longer cycles are rare but do crop up.

We can begin with any number, find the sum of its proper divisors, and repeat the process, forming what is known as the number's *aliquot sequence*. The result is often a little disappointing in that typically we get a chain that heads to 1 quite rapidly, at which point the process stalls. For example, even beginning with a promising-looking number such as 12, the chain is short:

$$12 \to (1 + 2 + 3 + 4 + 6) = 16 \to (1 + 2 + 4 + 8) = 15 \to (1 + 3 + 5) = 9$$

$$9 \to (1 + 3) = 4 \to (1 + 2) = 3 \to 1.$$

The trouble is, once you hit a prime, you are finished. The perfect numbers are of course exceptions, each giving us a little loop, while an amicable pair leads to a two-cycle: $220 \to 284 \to 220 \to \cdots$. Numbers that lead to cycles longer than two are called *sociable*. They were not studied at all until the 20th century as no one had ever found any. Even today, no number that leads to a three-cycle has been discovered, although there are now 120 known cycles of length four. The first examples were found by P. Poulet in 1918. The first is a five-cycle:

$$12,496 \to 14,288 \to 15,472 \to 14,536 \to 14,264 \to 12,496.$$

Poulet's second example is quite stunning, and to this day no other cycle has been found that comes close to matching it: starting with 14,316 we obtain a cycle of length 28. All other known cycles have length less than 10. To the present day, there are no theorems on amicable and sociable numbers as beautiful as those of Euclid and Euler on perfect numbers. However, modern computing power has led to something of an experimental renaissance in this kind of topic and there is more that can be said.

We can divide all numbers into three types, *deficient, perfect*, and *abundant* according to whether the sum of their proper divisors is less than, is equal to, or exceeds the number itself. For example, as we have already seen, 12 is an abundant number, as are 18 and 24 as the respective sums of their proper divisors are 21 and 36.

A naive search for abundance among the integers might lead you to guess that the abundant numbers are simply the multiples of 6. Certainly, any number greater than 6 of the form $6n$ is abundant, as the factors of $6n$ must include 1, 2, 3 together with n, $2n$, and $3n$, which sum to more than the original number $6n$. This observation can be extended, however, to show that abundance is not just about sixes as we can argue the same way for any perfect number k. The factors of nk will include 1 together with all the factors of the perfect number k, each multiplied by n so that the sum of all the proper factors of nk will add up to at least $1 + nk$, and therefore any multiple of a perfect number will be abundant. For example, 28 is perfect and hence $2 \times 28 = 56, 3 \times 28 = 84$ etc. are all abundant.

And so we see that multiples of perfect numbers and indeed, by the same token, multiples of abundant numbers are themselves abundant. Having made this discovery, you still might guess that all abundant numbers are simply multiples of perfect numbers. However, you don't have to look too much further to find the first exception to this conjecture, for 70 is abundant but none of its factors are perfect. Indeed, 70 is the first so-called weird number, but not exactly for this reason (the source of this label is explained below).

Despite these discoveries, you might still think it likely that, just as there seem to be no odd perfect numbers, there are no odd abundant numbers either. In other words, our modified conjecture might be that all odd numbers are deficient. Calculation of the aliquot sums of the first few hundred odd numbers would seem to confirm this theory, but the claim is eventually debunked upon testing 945, which has 975 as the sum of its proper divisors. Now the floodgates open as any multiple of an abundant number is abundant, and in particular the odd multiples of 945 immediately supply us with infinitely many more odd abundant numbers.

If we act a little more shrewdly, however, we can discover this counter-example more quickly than if we unthinkingly test one odd number after another. For a number to have a large aliquot sum, it needs lots of factors and large factors at that, which themselves come from being paired with small factors. We can therefore *build* numbers with large aliquot sums by multiplying small primes together. If we are focusing on odd numbers only, we should look at those that are products of the first few odd primes, which are 3, 5, 7, etc. This rule of thumb would soon lead you to test $3^3 \times 5 \times 7 = 945$ and thereby discover the abundance property among the odd numbers also.

It is not that unusual to find that the smallest example of a number with certain properties turns out to be rather large. This is especially true if the specified properties implicitly build a certain factor structure into the required numbers. The smallest example can then turn out to be gigantic, although not necessarily hard to find if we exploit the given properties in our quest for the solution. An example of a number riddle of this kind is to find the smallest number that is five times a cube and three times a fifth power. The answer is

$$7, 119, 140, 125 = 5 \times 1125^3 = 3 \times 75^5.$$

The reason why the smallest solution is in the billions, however, is not hard to see. Any solution n has to have the form $3^r 5^s m$ for some positive powers r and s and where the remaining prime factors are collected together into a single integer m that is not divisible by 3 or 5. If we first focus on the possible values of r, we observe that since n is 5 times *a cube*, the exponent r must be a multiple of 3, and since n is 3 times a *5th power*, the number $r - 1$ has to be a multiple of 5. The smallest r that satisfies both these conditions simultaneously is $r = 6$. In the same way, the exponent s has to be a multiple of 5, while $s - 1$ has to be a mutiple of 3 and the least s that fits the bill is $s = 10$. To make n as small as possible, we take $m = 1$ and so $n = 3^6 \times 5^{10} = 3(3 \times 5^2)^5 = 3 \times 75^5$, so that n is indeed 3 times a 5th power and at the same time $n = 5(3^2 \times 5^3)^3 = 5 \times 1125^3$, and so n is also 5 times a cube.

An even more extreme example is the celebrated Cattle Problem, attributed to Archimedes (287–212 BC), the greatest mathematician of antiquity. It was not solved until the 19th century. The smallest herd of cattle that satisfies all the imposed constraints in the original 44-line poem is represented by a number with over 200,000 digits!

A warning to be gleaned from all this is that numbers do not display their full variety until we move into the realms of the very large. For that reason, the mere fact that there are no odd perfect numbers with fewer than 300 digits does not in itself give grounds for saying that they 'probably' do not exist. All the same, it is the case that some leading experts in the field would be astonished if one ever turned up.

Returning once again to the general behaviour of aliquot sequences, there are still simple questions that may be put that no one can answer. What possibilities are open to aliquot sequences? If the sequence hits a prime, it will immediately terminate thereafter at 1, and cannot do this in any other way. If this does not happen, the sequence could be cyclic and so represent a sociable number. There is, however, another related possibility that is revealed by calculating the aliquot sequence of 95:

$$95 = 5 \times 19 \rightarrow (1 + 5 + 19) = 25 = 5 \times 5 \rightarrow (1 + 5) = 6 \rightarrow 6 \rightarrow 6 \rightarrow \cdots .$$

What has happened here is that although 95 is not itself a sociable number, its aliquot sequence eventually hits a sociable number (or more precisely in this case, the perfect number 6) and then goes into a cycle.

There is conceivably one possibility remaining, that being that the aliquot sequence of a number never hits a prime nor a sociable number, in which case the sequence must be an unending series of different numbers, none of which are either prime or sociable. Is this possible? Surprisingly, no one knows. What is more surprising is that there are small numbers whose aliquot sequence remain unknown (and thereby remain candidates for having such an infinite aliquot sequence). The first of these mysterious numbers is 276, whose sequence begins:

$$276 \to 396 \to 696 \to 1104 \to 1872 \to 3770 \to 3790 \to$$
$$\to 3050 \to 2716 \to 2772 \to \cdots$$

but no one knows exactly where it ends up.

It might well be that the reader would like to explore a little on their own, in which case I should let you in on the secret of how to calculate the so-called *aliquot function* $a(n)$ from the prime factorization of n – take the product of all terms $(p^{k+1} - 1)/(p - 1)$, where p^k is the highest prime power of the prime p that divides n, and then subtract n itself. For example, $276 = 2^2 \times 3 \times 23$ and so

$$a(276) = \frac{2^3 - 1}{2 - 1} \times \frac{3^2 - 1}{3 - 1} \times \frac{23^2 - 1}{23 - 1} - 276 =$$
$$7 \times 4 \times 24 - 276 = 672 - 276 = 396$$

as indicated by the second term in the aliquot sequence for 276 listed above.

There is no end to the types of numbers that we can introduce by giving a name to the numbers n that bear a certain relationship to the aliquot function. As we have already mentioned, n is *perfect* if $a(n) = n$ and *abundant* if $a(n) > n$. A *semiperfect number n* is one that is the sum of some of its proper divisors (those less than n), so it follows from the definition that all semiperfect numbers are either perfect or abundant. For example, 18 is semiperfect as $18 = 3 + 6 + 9$. A number is called *weird* if it is abundant but *not* semiperfect, and the smallest weird number is 70.

One can take the view that the topic is becoming too miscellaneous in nature – bestowing names on rather arbitrarily defined classes of numbers does not of its own accord make them interesting. We should know when to stop. That said, it is worth appreciating that the underlying strategies used to tackle these new questions are yet reminiscent of what Euclid and Euler showed us in relation to perfect numbers. You will recall that what Euclid proved was that *if* a Mersenne number was prime *then* another number was even and perfect. Euler then proved conversely that all even perfect numbers arise from this approach. In the 9th century, the Persian mathematician Thabit ibn Qurra introduced for any number n a triple of numbers which, *if all prime*, allowed the construction of an amicable pair. Thabit's construction was generalized further by Euler in the 18th century, but even this enhanced formulation only seems to yield a few amicable pairs and there are many amicable pairs that do *not* arise from this construction. (There are now nearly 12 million known pairs of amicable numbers.) In modern times, a similar approach by Kravitz gives a construction of weird numbers from certain numbers should they happen to be prime, and this formula has successfully found a very large weird number with more than fifty digits.

These last two chapters have served to familiarize the reader with factors and factorization of the natural numbers, or *positive integers* are they are also known, illustrated through a variety of examples. This will stand you in good stead for the upcoming chapter, in which you will learn how those ideas are applied to contemporary cryptography, the science of secrets.

Chapter 4
Cryptography: the secret life of primes

The reader will now appreciate that the collection of counting numbers has, from the earliest times, been recognized as the repository of riddles and secrets, many of which have never been revealed to this day. For many of us, this is enough to justify the continued serious study of numbers but others may take a different attitude. Intriguing and difficult as these conundrums may be, it might be imagined that they have little bearing on the rest of human wisdom. But that would be a mistake.

Over the last few decades it has emerged that ordinary secrets, of the kinds we all indulge in from time to time, can be coded as secrets about numbers. This has now all been put into practice and our most precious secrets, whether they be commercial or military, personal or financial, political or downright scandalous, can all be protected on the Internet by masking them using secrets about ordinary counting numbers.

Secrets turned into numbers

How is all this possible? Any information, whether it be a poem or a bank statement, a blueprint for a weapon or a computer program, can be described in words. We may, however, need to augment the alphabet that is used to make up our words beyond the ordinary letters of the alphabet. We may include number symbols, punctuation symbols including special symbols for space between ordinary words, but it is nonetheless the case that all the information we wish to transfer, including instructions for producing pictures and diagrams, can be expressed using words from an alphabet of, let us say, no more than one thousand symbols. We can count these symbols and so represent each symbol uniquely as a number. Since numbers are cheap and inexhaustible, it may be convenient to use numbers all with the same number of digits for this purpose (so, for example, every symbol was represented uniquely by its own four-digit PIN). We could string the symbols together as required to give one big long number that told the entire story. We can even work in binary if we wish and so devise a way of translating any information into one long string of 0s and 1s. Every message we might ever want to send could then be coded as a binary string and then decoded at the other end by a suitably programmed computer, to be compiled in ordinary language that we can all comprehend. This then is the first realization: in order to send messages between one person and another, it is enough, both in theory and in practice, to be able to send numbers from one person to another.

Turning messages into numbers, however, is not the big idea. To be sure, the exact process by which all the information is digitized may be hidden from the general public, but nonetheless is not the source of protection from eavesdroppers. Indeed, from the point of view of cryptography, we may identify any message, the so-called *plaintext*, with the number that represents it and thereby think of that number as the plaintext itself, as it is assumed that anyone has access to the wherewithal that will allow one to be transformed into the other. Secrecy only comes on to the scene when we mask these plaintext numbers with other numbers.

Let me introduce you to the fictitious characters that populate the various scenarios of cryptography, which is the study of *ciphers* (secret codes). We imagine Alice and Bob, who want to communicate with each other, without being overheard by the eavesdropper, Eve. Instinctively, we might sympathize with Alice and Bob, picturing Eve as up to no good, but of course the reverse may be true, with Eve representing a noble policing authority striving to protect us all from the evil plots of Bob and Alice.

Whatever the moral standing of the participants, there is an age-old approach that Alice and Bob may employ to cut Eve out of the conversation even if Eve intercepts messages that pass between them. They can encrypt the data using a cipher key that is known only to Alice and Bob. What they may arrange to do is to meet in a secure environment where they exchange with one another a secret number (let us say 57) and then return home. When the time comes, Alice will want to send a message to Bob and, just to illustrate the point, suppose that message can be represented by a single digit between 1 and 9. On the big day, Alice wants to send the message '8' to Bob. She takes her message and adds the secret ingredient, that is to say she masks its true value by adding 57 and so sends the message to Bob, across an insecure channel, of $8 + 57 = 65$. Bob receives this message and subtracts the secret number to retrieve Alice's plaintext $65 - 57 = 8$.

The nefarious Eve, however, has a good idea what these two are up to and indeed does manage to intercept the enciphered message, 65. But what can she do with it? She may know, as we do, that Alice has sent one of the nine possible messages $1, 2, 3, \cdots, 9$ to Bob and also knows that she has encrypted it by adding a number to the message, which must therefore lie between $65 - 9 = 55$ and $65 - 1 = 64$. However, because she cannot tell which of these nine masking numbers has been used (she isn't in on the secret), she is none the wiser as to the actual plaintext message that Alice sent to Bob, which is still just as likely to be any one of the nine possibilities. All she knows is that Alice has sent a message to Bob but has no idea what it is.

It might seem that Alice and Bob are now impervious to the malice of Eve and can communicate with impunity using the magic number 57 to disguise all they have to say. That, however, isn't quite the case. They would be well advised to change that number, indeed they are better off using a new secret number every time because if they don't, the system will begin to leak information to Eve. For example, say in a future week Alice wants to send to Bob the same message number 8. Everything would run as before and once again Eve would intercept the mysterious number 65 from the airwaves, but this time it would tell her something. Eve would know that, whatever this message is, it is the same message that Alice sent to Bob in the first week – this is just the sort of thing Alice and Bob would not want Eve to know.

This, however, looks to be no big problem for Alice and Bob. When they first meet up to 'exchange keys', instead of agreeing on just one secret number, Alice could provide Bob with a long ordered list of thousands of secret numbers, to be used one after another, thus avoiding the possibility of meaningful coincidences in their publicly available communications.

And this is indeed what is done in practice. This kind of cipher system is known in the trade as a *one-time pad*. The sender and receiver mask their plaintext with a single-use number from the 'pad'. That leaf of the pad is then discarded by both the sender and receiver after the message has been sent and deciphered. The one-time pad represents a completely secure system in that the insecure message that travels in the public domain contains no information about the content of the plaintext. To decipher it, the interceptor needs to get hold of that pad in order to obtain the encryption–decryption key.

Keys and key exchange

It would seem then that the problem of secure communication is completely solved by the one-time pad and, in a way, that is true. The difficulty with ciphers like the one-time pad, however, is that they require the participants to exchange a key in order to use them. In practice, this takes a lot of effort. For high-level communications, such as those between the White House and the Kremlin, money is no object and the necessary exchanges are carried out under conditions of maximum security. In the everyday world on the other hand, all sorts of people and institutions need to communicate with one another in a confidential fashion. The participants cannot afford the time and energy required to secure key exchanges and, even if this were arranged by a trusted third party, it can be an expensive business.

The common drawback of all ciphers that had been used for thousands of years up until the 1970s was that they were all *symmetric ciphers*, meaning that the encryption and decryption keys were essentially the same. Whether it was the simple alphabet-shift cipher of Julius Caesar, or the complex Enigma Cipher of the Second World War, they all suffered from the common weakness that once an adversary learned how you were encoding your messages, they could decode them just as well as you. In order to make use of a symmetric cipher, the communicating partners needed to exchange the cipher key in a secure way.

It seemed to have been tacitly assumed that this was an unavoidable principle of secret codes – for a cipher to be used the partners needed, somehow or other, to exchange the key to the cipher and to keep it secret from the enemy. Indeed, this might be regarded as mathematical common sense.

This is the kind of assumption that makes a mathematician suspicious. We are dealing with what is essentially a mathematical situation, so one would expect such a 'principle' to be well founded and represented by some form of mathematical theorem. Yet there was no such theorem, and the reason that there was no such theorem was that the principle simply is not valid, as the following thought experiment reveals.

Transmission of a secure message from Alice to Bob does not in itself necessitate the exchange of the key to a cipher, for they can proceed as follows. Alice writes her plaintext message for Bob, and places it in a box that she secures with her own padlock. Only Alice has the key to this lock. She then posts the box to Bob, who of course cannot open it. Bob, however, then adds a second padlock to the box, for which he alone possesses the key. The box is then returned to Alice, who then removes her own lock, and sends the box for a second time to Bob. This time, Bob may unlock the box and read Alice's message, secure in the knowledge that the meddling Eve could not have peeked at the contents during the delivery process. In this way, a secret message may be securely sent on an insecure channel without Alice and Bob ever exchanging keys. This imaginary scenario shows that there is no law that says that a key *must* change hands in the exchange of secure messages. In a real system, Alice and Bob's 'locks' might be their own coding of the message rather than a physical device separating the would-be eavesdropper from the plaintext. Alice and Bob may then use this initial exchange to set up an ordinary symmetric cipher that would be used to mask all their future communication.

Indeed, this is the way a secure communication channel is often established in the real world. Replacing physical locking devices by personal codes is not, however, so easy to do. Unlike the locks, the encodings of Alice and Bob may interfere with one another, making the unscrambling (that is, the unlocking) that is carried out first by Alice and then by Bob unworkable. However, that this method can be effective was first publicly demonstrated by Whitfield Diffie and Martin Hellman in 1976.

A second related approach is the idea of *asymmetric* or *public key cryptography* in which everyone publishes their own public key that is then used to encipher messages meant for that person. However, each person also holds a private key, without which the messages enciphered using their own unique public key cannot be read. In terms of the padlock metaphor, Alice provides Bob with a box in which to place his plaintext message together with an open padlock (her public key) to which she alone holds the key (her private key).

A workable public key system might seem too much to ask for as the twin requirements of security and ease of use seem to conflict. Fast, safe encryption is, however, available to the general public on the Internet, even if they barely realize that it is there, safe-guarding their interests. And it is all down to numbers, and prime numbers at that.

How secret primes protect our secrets

Remember that every plaintext message is regarded just as a single number, so it is natural to try to mask this number using other numbers. The most common way to do this is through employing the so-called RSA enciphering process, published in 1978 by its founders, Ron Rivest, Adi Shamir, and Leonard Adleman. In RSA, each person's private key consists of three numbers, p, q, and d, where p and q are (very large) prime numbers and the third ingredient d is Alice's secret deciphering number, the role of which will be explained in due course. Alice provides the public with $n = pq$, the product of her two secret primes, and an enciphering number e (which is an ordinary whole number, in no way related to the special constant called e mentioned in Chapter 2).

A simple example for the purposes of illustration would be for Alice to have the primes $p = 5$ and $q = 13$ so that $n = 5 \times 13 = 65$. If Alice sets her enciphering number to be $e = 11$, then her public key would be $(n, e) = (65, 11)$. To encrypt a message m, Bob only needs n and e. However, to decipher the encrypted message $E(m)$ that Bob transmits to Alice requires the deciphering number d, which in this case turns out to be $d = 35$, as we shall show a little further on. The mathematics that allows d to be calculated requires that the primes p and q are known. In this toy example, given that $n = 65$, anyone would soon discover that $p = 5$ and $q = 13$. However, if the primes p and q are *extremely* large (typically they are hundreds of digits in length), this task becomes a practical impossibility for almost any computer system, at least in a reasonably short time, such as two or three weeks. In summary, the RSA system of enciphering is based on the empirical fact that it is prohibitively difficult to find the prime factors of a very, very large number n. The clever part, which we shall explain in the remainder of the chapter, lies in devising a way that the message number m can be enciphered just using the publicly known numbers n and e but, in practice, deciphering requires possession of the prime factors of n.

Here is how it works. What Bob sends through the ether is not m itself but *the remainder* when m^e is divided by n. Alice can then recover m by taking this remainder r and similarly calculating the remainder when r^d is divided by n. The underlying mathematics ensures that the outcome for Alice is the original message m, which can then be decoded into ordinary plaintext by Alice's computer system. This is, of course, happening seamlessly behind the scenes for any real-life Alice and Bob.

It would seem that the only thing that Eve lacks that really matters is this deciphering number d. If Eve knew that, she could decipher the message just as well as Alice. It turns out that d is a solution of a certain equation. Solving this equation is computationally quite easy and relies on the Euclidean Algorithm, published in the Books of Euclid in 300 BC. That is not the difficulty. The trouble is that it is not possible to find out exactly what equation to solve unless you know at least one of the primes p and q, and that is the obstacle that stops Eve in her tracks.

We can explain more about how the numbers involved in all this work into the system. First, there is apparently quite a problem with Bob's initial task. The number m is big, the number n is monstrous (of the order of 200 digits) and even if e is not that large, the number m^e is going to be extremely large as well. After calculating it, we have to divide m^e by the number n to get the remainder r, which represents the enciphered text. It might seem that the calculations are too unwieldy to be practical. We should be aware that even though modern computers are extremely powerful, they yet have their limitations. When calculations involve very high powers, they can exceed the capacity of *any* computer system. We certainly cannot assume that any practical calculation that we set for a computer can be done in a short period of time.

The saving grace for Bob is that it is possible to find the required remainder r, without doing the long division at all. Indeed, the remainders just depend on remainders, and here is an example to illustrate the point. What are the final two digits of 7^{39}? (That is to say, what is the remainder when this number is divided by 100?) In order to answer this question, we might begin by calculating the first few powers of 7: $7^1 = 7$, $7^2 = 49$, $7^3 = 343$, $7^4 = 2,401$, $7^5 = 16,807$, $\cdots$. It will soon become clear, however, that the sheer size of these numbers is going to become unmanageable well before we get anywhere near 7^{39}. On the other hand, as we calculate one power after another, we see a pattern emerging. The key observation is that, as we calculate succeeding powers, the final two digits of the answer depend only on the final two digits of the preceding number, as when we carry out the multiplication, digits in the hundreds column and beyond have no effect on what end up in the units and tens columns.

What is more, since 7^4 has 01 as the final digit pair, the next four powers will end in 07, 49, 43, and then 01 again. Hence, as we compute succeeding powers, the pattern of the last two digits will simply repeat this cycle of length four, over and over again. To return to the question in hand, since $39 = 4 \times 9 + 3$, we will pass through this four-cycle nine times and then take three more steps in calculating the final two digits of 7^{39}, which must therefore be 43.

And this works quite generally. In order to find the remainder when some power a^b is divided by n say, we need only take the remainder r when a is divided by n and keep track of the remainders as we take successive powers of r. When we work with the remainder r, which will be a number in the range from 0 to $n - 1$, mathematicians say that we are working *modulo n*, discarding any higher multiples of n that may arise, as they leave a remainder of 0 when divided by n, and so cannot contribute to the value of the final remainder r.

You might still suspect that I have rigged the evidence by choosing an example where a very small power left a remainder of 1 – in this instance 7^4 was 1 more than a multiple of $n = 100$. This, however, is only partly true. It turns out that, if we take any two numbers, a and n, whose highest common factor is 1 (we say such numbers are mutually *coprime*), then there is always a power t such that a^t equals 1 modulo n, that is to say leaves a remainder of 1 when divided by n. From this point, the remainders of successive powers follow a cycle of length t. It can, however, be hard to predict what is the value of t, but it is known that t must always equal or be a factor of a number traditionally written as $\phi(n)$, the value of the Euler *phi function*.

And so what is $\phi(n)$? It is defined as the count of the numbers up to n that are coprime with n. For example, if $n = 15$, then the set of numbers in question is {1, 2, 4, 7, 8, 11, 13, 14}, which lists all the numbers up to 15 that have no factor in common with 15 (except the inevitable factor of 1). Since this set has 8 members, we see that $\phi(15) = 8$. Fortunately there is a slicker way of finding $\phi(n)$ that does not entail explicitly listing all the integers coprime with n and then counting them up. As with most functions of this nature, the value can be expressed in terms of the prime factorization of n. Indeed, we only need know the prime factors of n, for $\phi(n)$ can be found by taking n and multiplying it by each of the fractions $(1 - \frac{1}{p})$ as p ranges over all the prime divisors of n. The prime factors of 15, for instance, are 3 and 5, so the answer in this case is

$$\phi(15) = 15 \times (1 - \frac{1}{3})(1 - \frac{1}{5}) = 15 \times \frac{2}{3} \times \frac{4}{5} = 8,$$

which is the same as the result that we obtained directly from the definition. Using this method, you might like to check yourself that $\phi(100) = 40$, and so, for instance, it then follows that 7^{40} equals 1 modulo 100. However, as we have already seen, the least power of 7 that yields a remainder of 1 is not 40 but its divisor 4.

All this serves to give an indication that the number sent by Bob to Alice, m^e modulo n, can indeed be calculated without too much effort on behalf of Bob's computer. All the same, the numbers involved are in practice mighty big, so more explanation is needed to show that they can be handled. The large powers involved in computing m^e can be dealt with in stages by a process known as *fast exponentiation*. Without going into detail, the method involves successive squaring and multiplying of powers to arrive at m^e modulo n with the binary form of e guiding the algorithm through to quickly find the required remainder in relatively few steps.

Euclid shows Alice how to find her deciphering number

The deciphering number is the receiver's magic wand that allows the message retrieval. This number d is chosen so that the product de leaves a remainder of 1 when divided by $\phi(n)$. Since $n = pq$ is a product of two distinct primes, the value of $\phi(n) = pq(1 - \frac{1}{p})(1 - \frac{1}{q}) = (p - 1)(q - 1)$. It turns out that there is always just one value for d in the range up to $\phi(n)$ that has the required property.

Alice's computer can find d using an algebraic tool that is over 2,300 years old, the Euclidean Algorithm, which will be explained in a moment. Eve's computer could of course do the same thing if it just knew which equation to solve. However, since p and q are private to Alice, so is $(p-1)(q-1)$ and Eve does not know where to begin.

The existence of d is only guaranteed if a certain mild restriction is placed on the (publicly declared) enciphering number e. Alice must ensure that e has no common prime factor with $\phi(n)$. This is quite easily done as Alice can test $\phi(n)$ for division by particular primes and so ensure that e meets these requirements without compromising the identity of p and of q. Indeed, the value of e often used in practice is the fourth so-called *Fermat prime* $e = 65537 = 2^{16} + 1$; this value, $2^{2^4} + 1$ has a particularly rare property, that being that it is possible to construct a regular polygon with e sides using a straightedge and compass. Its utility in cryptography, however, is due to it being a fairly large prime that exceeds a power of 2 by exactly 1, which lends well to the fast exponentiation process mentioned earlier.

Returning to the Euclidean Algorithm, this begins from the observation that it is possible to find the *highest common factor* (hcf) of two numbers $a > b$ by successive subtraction. (The hcf is also known as the gcd – *greatest common divisor*.) We just note that $r = a - b$ has the property that any common factor of any two of the three numbers a, b, and r will also be a factor of the third. For example, if c is a common factor of a and b, so that $a = ca_1$ and $b = cb_1$ say, we see that $r = a - b = ca_1 - cb_1 = c(a_1 - b_1)$, giving us a factorization of r involving the divisor c. In particular, the hcf of a and b is the same as the hcf of b and r. Since both these numbers are less than a, we now have the same problem but applied to a smaller number pair. Repetition of this idea then will eventually lead to a pair where the hcf is obvious. (Indeed, the two numbers in hand will eventually be the same, for if not we could proceed one more step; their common value is then the number we seek.)

For example, to find the hcf of $a = 558$ and $b = 396$, the first subtraction would give us $r = 558 - 396 = 162$, so our new pair would be 396 and 162. Since $396 - 162 = 234$, our third pair becomes 234 and 162, and as we continue the full list of number pairs is:

$$(558, 396) \to (396, 162) \to (234, 162) \to (162, 72) \to (90, 72) \to$$
$$\to (72, 18) \to (54, 18) \to (36, 18) \to (18, 18)$$

and so the hcf of 558 and 396 is 18.

It is possible to write down the hcf of a number pair from the prime factorizations of the numbers in question. In this example, $558 = 2 \times 3^2 \times 31$, while $396 = 2^2 \times 3^2 \times 11$; taking the common power of each prime entering into the factorizations, we obtain the hcf as $2 \times 3^2 = 18$. Nevertheless, for larger numbers it takes much less work to use Euclid's Algorithm as it is generally easier to perform subtractions than to find prime factorizations.

Another bonus of the Euclidean Algorithm is that it is always possible to work it backwards and in so doing express the hcf in terms of the original two numbers. To see this in action in the previous example, it is best to compress the calculation when the same number appears several times over in the course of the subtractions, representing this as a single equation as follows:

$$558 = 396 + 162$$
$$396 = 2 \times 162 + 72$$
$$162 = 2 \times 72 + 18$$
$$72 = 4 \times 18.$$

Beginning with the second to last line, we now use each little equation to eliminate the intermediate remainders, one at a time. In this example, by using first the penultimate equation, and then the one above that we obtain:

$$18 = 162 - 2 \times 72 = 162 - 2 \times (396 - 2 \times 162) = 5 \times 162 - 2 \times 396$$

and finally using the first equation we can eliminate the first intermediate remainder of 162:

$$= 5 \times (558 - 396) - 2 \times 396 = 5 \times 558 - 7 \times 396 = 18.$$

That we can perform this reverse procedure is important for both practical and theoretical reasons. In particular, to find Alice's deciphering number d, we want d to satisfy the condition that de leaves a remainder of 1 when divided by $\phi(n)$. (For brevity, we shall denote $\phi(n)$ by the single symbol k.) We can now see the reason why we insist on e and k being a coprime pair, as if their highest common factor is 1, when we act the Euclidean Algorithm on the pair e and k, the final remainder that appears is, of course, 1. By reversing the algorithm, we will eventually express 1 as a combination of e and k; in particular, we will find integers c and d such that $ck + de = 1$, or in other words $de = 1 - ck$, so that de will leave a remainder of 1 when divided by k.

This relatively simple process will yield Alice's deciphering number d: the initial value of d obtained from the equation may not lie in the range from 1 to k but if not, by adding or subtracting a suitable multiple of k, we will eventually find the unique number d in that range that has the magic property that de leaves a remainder of 1 when divided by k. (The uniqueness of d is easily proved, but we won't digress into further explanation here.) That is how the deciperhing number d is calculated as we can show by returning to the example given earlier where $p = 5$, $q = 13$, so that $n = pq = 5 \times 13 = 65$. We have $\phi(n) = (p - 1)(q - 1) = 4 \times 12 = 48$. Alice sets $e = 11$, and since 11 and 48 are coprime, this is within the rules of the game. The Euclidean Algorithm applied to $\phi(n) = k = 48$ and $e = 11$ then gives:

$$48 = 4 \times 11 + 4$$
$$11 = 2 \times 4 + 3$$
$$4 = 1 \times 3 + 1$$

confirming that the hcf of k and e is indeed 1. Reversing the algorithm we obtain:

$$1 = 4 - 3 = 4 - (11 - 2 \times 4) = 3 \times 4 - 11 = 3(48 - 4 \times 11) - 11$$
$$= 3 \times 48 - 13 \times 11.$$

This gives an initial value of $d = -13$ as the solution to the requirement that $11d$ leaves remainder 1 upon division by 48, so in order to get a positive value of d in the required range we add 48 to this number to get $d = 48 - 13 = 35$.

The reason why d works for Alice is all down to modular arithmetic and the fact that de leaves a remainder of 1 when divided by $k = \phi(n)$. Alice calculates $(m^e)^d = m^{de}$ modulo n. Now de has the form $1 + kr$ for some integer r. As explained before, m^k leaves a remainder of 1 when divided by n (this is often known as *Euler's Theorem*) and so the same is true of $(m^k)^r = m^{kr}$. Hence $m^{1+kr} = m \times m^{kr}$ leaves the remainder m when divided by n. (Detailed verification of this requires a little algebra, but that is what happens.) In this way, Alice retrieves Bob's message, m.

And in passing it is well to point out that the Euclidean Algorithm provides the missing link in our proof of the uniqueness of prime factorization as it allows us to verify the euclidean property that if a prime p is a factor of the product ab, so that $ab = pc$ say, then p is a factor of at least one of a and b. The reason for this is that if p is *not* a factor of a then, since p is prime, the hcf of a and p is 1. By reversing the Euclidean Algorithm when applied to the pair a and p, we can then find integers r and s say such that $ra + sp = 1$. This is enough to show that p is then a factor of b for, since $ab = pc$, we have:

$$b = b \times 1 = b(ra + sp) = r(ab) + psb = r(pc) + psb = p(rc + sb).$$

This is the required factorization of b that features the prime p as a factor.

In conclusion, the number theory underlying RSA enciphering makes the system sound, although various protocols that have not been explained here must be respected in order to safeguard the integrity of the system. There are issues of *authentification* (what if Eve contacts Alice pretending to be Bob?), *non-repudiation* (what if Bob pretends that it was Eve who sent his message to Alice?), and *identity fraud* (what if Alice abuses confidential identification sent to her by Bob and tries to impersonate him online?). Moreover, other weaknesses in the system can be exposed when predictable or repeated messages proliferate. However, these difficulties may potentially arise in any public key cryptosystem. They can be overcome and in the main are unrelated to the underlying mathematical techniques that ensure high quality and robust encyryption.

This chapter has demonstrated a major application of prime numbers and the theory of divisibility and remainders. The ancient mathematics of Euclid and the 18th-century contribution of Euler allows this to be explained, not only in broad principle, but in fine detail.

The first part of our book closes with Chapter 5 which introduces some special classes of integers associated with the enumeration of some naturally occurring groupings.

Chapter 5
Numbers that count

Numbers that arise of their own accord in counting problems are important and so have been extensively investigated. Here I will describe the binomial coefficients, and the numbers of Catalan, Fibonacci, and Stirling because they enumerate certain natural collections. But we first begin with some very fundamental number sequences.

Triangular numbers, arithmetic and geometric progressions

Since they will reappear when we look at binomial coefficients, I will take a moment to revisit the *triangular numbers*, the nth of which, denoted by t_n, is defined as the sum of the first n counting numbers. Its value, in terms of n, can be found by the following trick. We write t_n as the sum just mentioned and then again as the same sum but in the reverse order. Adding the two versions of t_n together:

$$t_n = 1 + 2 + 3 + \cdots + (n-2) + (n-1) + n$$
$$t_n = n + (n-1) + (n-2) + \cdots + 3 + 2 + 1;$$

the outcome is of course an expression for $2t_n$. However, the point of doing this is that we have paired 1 with n, 2 with $n-1$, 3 with $n-2$, etc. The sum of each of these pairs is the same value, $n+1$, and there are n pairs altogether. In conclusion, we infer that

$2t_n = n(n + 1)$, or in other words, the value of the nth triangular number is $\frac{1}{2}n(n + 1)$. For example, the sum of all the integers from 1 up to 1000 is therefore equal to $500 \times 1001 = 500,500$.

Indeed, this formula allows us to find the rule for summing the first n terms of any *arithmetic* series, or *progression* as they are known, which is one of the form $a, a + b, a + 2b, a + 3b, \cdots$. We first take care of the change of scale. By multiplying through the expression for t_n by b we see that $b + 2b + 3b + \cdots + nb = \frac{b}{2}n(n + 1)$. To find the formula for the sum of the general arithmetic series, first note that the sum of the first $n - 1$ terms of the previous series is obtained by replacing n by $n - 1$ in the previous formula, to give $\frac{b}{2}n(n - 1)$. The general arithmetic series now comes from adding a to every term and including a as the first term as well. This means that we need to add na to the previous sum to give us the general formula for the sum of the first n terms of an arithmetic series:

$$a + (a + b) + (a + 2b) + \cdots + (a + (n - 1)b) = na + \frac{b}{2}n(n - 1).$$

For example, by taking $a = 1$ and $b = 2$, we see that the sum of the first n odd numbers is $n + n(n - 1) = n + n^2 - n = n^2$, the nth square.

If we replace addition by multiplication as the operation, we move from arithmetic series to *geometric series*. In an arithmetic series, each pair of successive terms is separated by a *common difference*, the number b in our notation. In other words, to move from one term to the next, we simply *add b*. In a geometric series, we once again begin with some arbitrary number, a as the first term and move from one term to the next by *multiplying* by a fixed number, called the *common ratio*, denoted by the symbol r. That is to say, the typical geometric series has the form $a, ar, ar^2, \cdots$ with the nth term being ar^{n-1}. As with arithmetic series, there is a formula for the sum of the first n terms of a geometric series:

$$a + ar + ar^2 + \cdots + ar^{n-1} = \frac{a(r^n - 1)}{r - 1}.$$

The quick way of seeing that this formula is right is to take the equivalent form that we obtain when we multiply both sides of this equation by $(r - 1)$ and multiply out the brackets. On the left-hand side we obtain:

$$(ar + ar^2 + ar^3 + \cdots + ar^n) - (a + ar + ar^2 + \cdots + ar^{n-1})$$

and the whole expression *telescopes*, meaning that nearly every term is cancelled by one in the other bracket: the only exceptions are $ar^n - a = a(r^n - 1)$, showing that our formula for the sum is correct. For example, putting $a = 1$ and $r = 2$ gives us the sum of powers of 2:

$$1 + 2 + 4 + \cdots + 2^{n-1} = 2^n - 1.$$

This formula is just what you need in order to verify Euclid's result from Chapter 3 on how to generate even perfect numbers from Mersenne primes.

Factorials, permutations, and binomial coefficients

As we have seen, the nth triangular number arises from summing all the numbers from 1 up to n together. If we replace addition by multiplication in this idea, we get what are known as the *factorial* numbers, which made their first appearance in Chapter 2.

Factorials come up constantly in counting and probability problems such as the chances of being dealt a certain type of hand in a card game like poker. For that reason, they have their own notation: the nth factorial is denoted by $n! = n \times (n - 1) \times \cdots \times 2 \times 1$. The triangular numbers grow reasonably quickly, at about half the rate of the squares, but the factorials grow much faster and soon pass into the millions and millions: for example $10! = 3,628,800$. The exclamation mark alerts us to this rather alarming rate of growth.

In particular, $n!$ is the number of recognizably different ways that you can arrange n objects, such as n numbered balls, in a row. This follows as you have n choices for which ball goes first, and for each such choice you have $n - 1$ balls remaining for the second place, $n - 2$ for the third, and so on. If we stop after selecting just r balls, and denote the corresponding number by $P(n, r)$, we see that there are $n \times (n - 1) \times (n - 2) \times \cdots \times (n - r + 1)$ *permutations*, as we say, of n balls, taking r at a time. This is also conveniently expressed as the ratio of two factorials: $P(n, r) = \frac{n!}{(n-r)!}$. For example, if you are dealt a poker hand, you pick up 5 cards from a deck of 52 and the number of ways this can happen is $P(52, 5) = 52 \times 51 \times 50 \times 49 \times 48$. However, a hand in a card game does not depend on the order in which you pick up the cards but only the collection of cards itself. For a poker hand, each set of 5 can itself be rearranged in $5! = 120$ ways so that the number of genuinely different 5-card hands is $P(52, 5)/120 = 2,598,960$, about two and a half million.

The most special class that emerges in counting problems, or *enumerations* as they are called, is that of the *binomial coefficients*, so named as they arise as the multipliers of powers of x when the binomial expression $(1 + x)^n$ is expanded. The binomial coefficient $C(n, r)$ is the number of different ways we may construct a set of size r from one of size n. For example, $C(4, 2) = 6$, as there are six pairs (taken without regard to order within a pair) that can be chosen from a group of four: for example, if we have four children, Alex, Barbara, Caroline, and David, there are six ways that we can select a pair from this group: AB, AC, AD, BC, BD, and CD.

The binomial coefficients can be calculated in two distinct ways. First, we can extend the argument above that we used to calculate $C(52, 5)$: in general we see that $C(n, r) = P(n, r)/r!$, which in turns gives us the useful expression:

$$C(n, r) = \frac{n!}{(n - r)!r!}.$$

A notable special case is when we let $r = 2$, which corresponds to the number of pairs that can be selected from a set of n objects.

The answer is $\frac{n!}{(n-2)!2!}$. Now all the factors in the $(n-2)!$ term in the denominator cancel with the corresponding factors in $n!$ and since $2! = 2$, the expression for $C(n, 2)$ simplifies to $\frac{n(n-1)}{2}$. In other words, the number of ways of choosing a pair from a collection of n objects is t_{n-1}, the $(n-1)$st triangular number. For instance, as we have already seen, $C(4, 2) = 6$, which is indeed the third triangular number.

This factorial-based formula for calculating binomial coefficients does give a nice algebraic hold on the binomial coefficients that allows us to demonstrate their many special properties. However, the evolution of these properties is often more transparent if we focus on a second way to generate these integers, which is by means of the *Arithmetic Triangle* (see Figure 2), also known as *Pascal's Triangle*, in honour of the 17th-century French mathematician and philosopher Blaise Pascal (1623–62). (The Arithmetic Triangle has been discovered and re-discovered throughout Persia, India, and China over the last 1,000 years: for example, it featured as the front cover of *The Precious Mirror* by Chu Shih-Chieh in 1303.)

Each number in the body of the triangle is the sum of the two above it. The triangle, which can be continued indefinitely, gives the full list of binomial coefficients. For example, to find the number of ways of selecting five people from a group of seven, proceed as follows. Number the lines of the triangle, beginning

```
                            1
                         1     1
                      1     2     1
                   1     3     3     1
                1     4     6     4     1
             1     5    10    10     5     1
          1     6    15    20    15     6     1
       1     7    21    35    35    21     7     1
    1     8    28    56    70    56    28     8     1
                            ⋮
```

2. The Arithmetic Triangle

with 0 at the top. Similarly number the positions within each row from left to right, again starting with 0. Go down to the line numbered 7, and then go to the number on that line numbered 5 (remembering to start your count from 0): we see the answer is 21. You will note the symmetry of each row: for example, 21 is also the number of ways of choosing two people from a group of seven. This is explained by observing that when we choose the five from seven, we are simultaneously choosing two from seven as well – the two being the pair left behind. This symmetry argument of course applies to every row. This is also manifested in the formula on page 55, for it returns the same expression if we replace r by $n - r$, as the terms r and $n - r$ that we see in the denominator simply swap positions.

The reason that the pattern gives the right answers is not hard to see. Each row builds from the one above it. We can see easily that the first three rows are correct: for example, the 2 in the centre of the third row tells us that there are two ways of choosing a single person from a pair. The 1 that sits on top is saying that there is one way to choose a set of size zero from the empty set. In fact, there is one way of choosing a set of size zero from any set, which is why every row begins with 1. Let us focus on the example just given – there are 21 = 15 + 6 ways of selecting five from a group of seven people. The 21 quintets naturally split into two types. First, there are 15 ways to form a group of four from the first six people, to which we may add the seventh person to form our fivesome. If we don't include the seventh person, however, then we have to build a set of five from the first six, and there are six ways of doing this. This illustrates how one row leads to the next: each entry is the sum of the two above it, and this pattern propagates throughout the triangle. In symbols this rule takes the form:

$$C(n, r) = C(n - 1, r) + C(n - 1, r - 1).$$

The triangle is rich in patterns. For example, summing all the numbers in each successive row gives the doubling sequence

1, 2, 4, 8, 16, 32, ⋯ : the sequence of powers of 2. In summing the row that begins 1, 8, 28, 56, ⋯ for instance, we are summing the number of ways of choosing a set of size 0, 1, 2, 3 etc. from a set of 8. In total, this gives us the number of ways of selecting a set of *any* size from a group of 8, which is equal to 2^8 as, in general, a set of size n contains 2^n subsets within it.

This last fact can be seen directly, for a subset of a set of size n can be identified by a binary string of length n in the following way. We consider the set in question in a specific order $\{a_1, a_2, \cdots, a_n\}$ say, and then a binary string of length n specifies a subset by saying that each instance of 1 in the string indicates the presence of the corresponding a_i in the subset in question. For example, if $n = 4$, the strings 0111 and 0000 stand respectively for $\{a_2, a_3, a_4\}$, and for the empty set. Since there are two choices for each entry in the binary string, there are 2^n such strings in all and therefore 2^n subsets within a set of size n.

Catalan numbers

Every second row in the Arithmetic Triangle has a number sitting in the middle: 1, 2, 6, 20, 70, 252, 924, ⋯. These numbers are divisible by the consecutive counting numbers 1, 2, 3, 4, 5, 6, 7, ⋯ and the numbers that come about as we carry out these divisions, 1, 1, 2, 5, 14, 42, 132 ⋯ are known as the *Catalan numbers*. In terms of these *central binomial coefficients*, the nth Catalan number can be expressed as $\frac{1}{n+1} C(2n, n)$ for $n = 0, 1, 2, \cdots$.

One of the simplest visual representations that gives rise to this number type is as the number of ways we can draw 'mountains' using n up strokes and n down strokes (see Figure 3)

Each mountain pattern has an interpretation, however, as a meaningful bracketing and so the number of meaningful ways of arranging a collection of n pairs of parentheses is the nth Catalan number. For example, $(())()$ and $((()))$ are meaningful bracketings

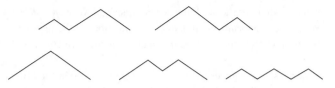

3. With three up and down strokes there are five mountain patterns

but ()(() is not: to be meaningful, the number of left brackets must never fall behind the number of right brackets as we count from left to right. This corresponds to the natural condition that our mountains must never dive underground. For instance, the first and last mountain patterns in Figure 3 correspond to the bracketing ()(()) and ()()() respectively.

The nth Catalan number also counts the number of ways that we can break up a regular polygon with $n + 2$ sides into triangles by means of diagonals that do not cross one another, and there are other interpretations along these lines. As with binomial coefficients, there are formulas relating Catalan numbers to smaller Catalan numbers, which makes them amenable to manipulation.

Fibonacci numbers

The Fibonacci sequence is one series of numbers that engenders wide fascination among the general public. The sequence runs as follows

$$1, 1, 2, 3, 5, 8, 13, 21, 34, 55, 89, 144, 233, 377, 610, \cdots$$

where each number after the pair of initial 1s is the sum of the two that come before. In this, there is a similarity with the binomial coefficients in that each term is the sum of two previous ones in the sequence, but the method of formation of the Fibonacci numbers is simpler:

$$f_n = f_{n-1} + f_{n-2}$$

where f_n denotes the nth Fibonacci number and we fix $f_1 = f_2 = 1$. We call such a formula that defines each member of a sequence in terms of its predecessors a *recursion* or a *recurrence relation*.

How does this sequence arise? It was first introduced in 1202 by Leonardo of Pisa, better known as Fibonacci, in the form of his celebrated Rabbit Problem. A female rabbit is born and after two months reaches maturity and thereafter gives birth to a daughter each month. The number of female rabbits we have at the beginning of each month is then given by the Fibonacci numbers, for there is 1 rabbit at the beginning of the first month, and the second, but at the start of the third month she gives birth to a daughter so we then have 2 rabbits. Next month she has another, giving 3 and the month after that we have 5 bunnies as both mother and her eldest daughter are now old enough to breed. In general, at the beginning of each month thereafter, the number of *newborn* daughters equals the number of females we had *two* months ago, as only they are old enough to breed. It follows that the number of females we have at the start of each subsequent month equals the total of the previous month (Fibonacci's rabbits are immortal) plus the number we had the month before that. Therefore the rule of formation of the Fibonacci numbers exactly matches the breeding pattern of his rabbits.

Despite the fact that real rabbits do not breed in this contrived fashion, Fibonacci numbers arise in nature in a variety of ways, including plant growth. The reasons for this are well understood but are related to more subtle attributes of the sequence connected to the so-called *Golden Ratio*, a number that we are about to introduce.

The simplest types of number progressions are the arithmetic and geometric progressions introduced in the first section. Although the Fibonacci sequence is neither of these, it does however have a surprising link with the latter type. If we form the sequence of differences of the Fibonacci sequence, because of the way the sequence is defined, we get $0, 1, 1, 2, 3, 5, 8, 13, \cdots$, that is we recover the Fibonacci sequence again except this time beginning at 0. This happens precisely because of the way the sequence is formed: the difference of two consecutive Fibonacci numbers is the one immediately preceding both in the sequence. (To see this algebraically, subtract f_{n-1} from both sides of the Fibonacci recurrence above.) Nor is the sequence a geometric progression as the ratio of consecutive Fibonacci numbers is not constant. All the same, when we look at the ratio of successive terms we see that it does seem to settle down to a limiting value. This near stable behaviour of the ratio comes about quite quickly, as we see as we divide each Fibonacci number by its predecessor:

$$\frac{34}{21} = 1 \cdot 6190, \ \frac{55}{34} = 1 \cdot 6176, \ \frac{89}{55} = 1 \cdot 6182, \ \frac{144}{89} = 1 \cdot 6180, \cdots$$

But what is the mysterious number, $1.6180\ldots$, which we see emerging? This number τ is known as the *Golden Ratio*, and it arises quite of its own accord in geometrical settings that look a world away from Fibonacci's rabbits. For example, τ is the ratio of the diagonal of a regular pentagon to its side (see Figure 4). Each diagonal meets another at a point that divides each into two parts that are themselves in the ratio $\tau : 1$. Pairs of intersecting sides and intersecting diagonals form the four sides of a rhombus (a 'square' parallelogram) *ABCD* as shown. Where diagonals cross, they form a smaller inverted pentagon.

A characteristic associated with the Golden Ratio is *self-similarity*, which means that objects such as the pentagon that involve τ often contain smaller copies of themselves within. This is seen in the rectangle with sides of lengths τ and 1, for it is unique in displaying the property that if we slice off the largest square we can (a square of side length 1) then the smaller rectangle that remains is a copy of the original. This figure is for that reason known as the *Golden Rectangle* (see Figure 4). The value of τ can be gleaned from the given property of the rectangle, for if we call the length of the longer side τ and make the shorter side one unit,

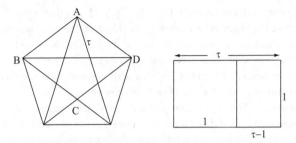

4. Pentagon and the Golden Rectangle

the similarity of the rectangles is captured by the equality $\frac{\tau}{1} = \frac{1}{\tau-1}$, which we obtain by equating the ratio of the longer to the shorter side of the rectangles. Cross-multiplying in this expression then yields the equation $\tau^2 - \tau = 1$. Using the standard formula to solve such a *quadratic equation* (one involving a square), we see that the positive root of this equation equals:

$$\tau = \frac{1 + \sqrt{5}}{2} = 1 \cdot 6180229 \cdots$$

Another way of retrieving this valuation of τ is through its so-called continued fraction, which ties τ directly to the Fibonacci numbers, and we shall explore this idea in Chapter 7.

In the long run, the Fibonacci sequence behaves like a geometric progression based on the Golden Ratio. It is this property, together with its simple rule of formation that causes the Fibonacci sequence to arise so persistently.

Stirling and Bell numbers

Like the binomial coefficients, the Stirling numbers often arise in counting problems and depend on two variables, n and r. The Stirling number $S(n, r)$ is the number of ways of partitioning a set of n members into r blocks (with no block empty, and the order of the blocks and within the blocks, is immaterial). (Strictly these are called Stirling numbers of the second kind. Those of the first kind, which are related, count something quite different, namely the number of ways we can permute n objects into r cycles.) For instance, the set with members a, b, c can be partitioned into three blocks in just one way: $\{a\}$, $\{b\}$, $\{c\}$, into two blocks in three ways $\{a, b\}$, $\{c\}$; $\{a\}$, $\{b, c\}$, and $\{a, c\}$, $\{b\}$, and into a single block in one way only: $\{a, b, c\}$; it follows that $S(3, 1) = 1$, $S(3, 2) = 3$ and $S(3, 3) = 1$. Since a set of n members can be partitioned in only one way into either 1 block or into n blocks, we always have $S(n, 1) = 1 = S(n, n)$. If we draw up the triangle of Stirling numbers after the fashion of Pascal's Triangle, we arrive at the array of Figure 5, and we now explain how the triangle is generated.

Once again, the numbers satisfy a recurrence relation, meaning that each can be related to earlier ones in the array. Indeed, as with the binomial coefficients, each Stirling number can be obtained from the two above it, but it is not simply the sum. What is more, the row symmetry we saw in the Arithmetic Triangle that generates the binomial coefficients is not present in Stirling's Triangle. For example, $S(5, 2) = 15$ but $S(5, 4) = 10$. The rule of recurrence is simple enough, however. The entry 90, for example, is equal to $15 + 3 \times 25$. This is indicative of the general situation: to find a number in the body of the triangle, take the two immediately above it, and add the first to the second *multiplied by the number of the position in the row you are at*. (This time, unlike the Arithmetic Triangle, start your row count at 1.) In a similar way, the entry $S(5, 4) = 10 = 6 + 4 \times 1$. It is only the part of the rule

```
                        1
                  1           1
            1           3           1
      1           7           6           1
1           15          25          10          1
1     31          90          65          15          1
1     63    301         350         140         21          1
                        ⋮
```

5. Stirling's Triangle

in italics that differs from that of the Arithmetic Triangle. That is enough, however, to make the study of Stirling numbers considerably more difficult to that of the binomial coefficients. For instance, we derived a simple explicit formula for each binomial coefficient in terms of the factorials. Similarly, there is a formula for the nth Fibonacci number in terms of powers of the Golden Ratio, but nothing of the kind exists for Stirling numbers.

The recurrence rule is not hard to explain. We argue similarly to that for the recursion for the binomial coefficients, and by doing so recover the recurrence outlined above that is identical in form except for a single multiplier. In order to form a partition of a set of size n into r non-empty blocks, we may proceed in two distinct ways. We may take the first $n - 1$ elements of the set and partition it into $r - 1$ non-empty blocks in $S(n - 1, r - 1)$ ways, and the final member of the set will then form the rth block. Alternatively, we may partition the first $n - 1$ elements of the set into r non-empty blocks, which can be done in $S(n - 1, r)$ ways, and then decide in which of the r blocks to place the final member of the set, giving a multiplier of r to that number. Hence we infer that

$$S(n, r) = S(n - 1, r - 1) + rS(n - 1, r) \text{ for } n = 2, 3, \cdots$$

Using this recursion formula, we may calculate each line of the Stirling Triangle from the one above it. For example, putting $n = 7$ and $r = 5$ we obtain:

$$S(7, 5) = S(6, 4) + 5S(6, 5) = 65 + 5 \times 15 = 65 + 75 = 140.$$

We can compute $S(n, 2)$ and $S(n, n - 1)$ directly from the definition as follows. An arbitrary partition of the n-set into a first set and a second set is described by a binary string of length n, where the presence of a 1 indicates presence in the first set and a 0 in the second (in a similar way to how we showed that the number of subsets of an n-set is 2^n). There are therefore 2^n such *ordered pairs* of sets. Since, however, there is no ordering of the blocks within a partition, we divide this number by 2 to find the number of partitions of the n-set into 2 sets, giving the number 2^{n-1}. Finally, we need to subtract 1 from this in order to exclude the case where one of the sets is empty; hence $S(n, 2) = 2^{n-1} - 1$. You can check that this represents the second diagonal line of numbers $1, 3, 7, 15, 31, 63, \cdots$ running from the top right to the bottom left in Figure 5.

At the other extreme, a partition of the n-set into $n - 1$ blocks is determined by a choice of the unique block of size 2. The number of ways of making this selection is $C(n, 2) = \frac{1}{2}n(n - 1)$, the $(n - 1)$st triangular number (see the second diagonal $1, 3, 6, 10, 15, 21, \cdots$ running from top left to bottom right in Figure 5).

The sum of any row of the Arithmetic Triangle gives the corresponding power of 2 –the number of subsets of a set of a given size. Similarly, summing the nth row of Stirling's Triangle gives the number of ways of breaking a set of n objects into blocks, and this is called the nth *Bell number*.

Partition numbers

If on the other hand, the n objects of the set to be partitioned are identical, and so cannot be distinguished from one another, the number of ways of splitting the whole collection up into blocks is a much smaller integer, known as the nth *partition number*. A particular partition corresponds to writing n as a sum of positive integers, without regard to order: for example, $1 + 1 + 1 + 1 + 1$ is one partition of 5 and there are six others, for we can also represent 5 as $1 + 1 + 1 + 2, 1 + 2 + 2, 1 + 1 + 3, 2 + 3, 1 + 4$, or simply as 5. Therefore the 5th partition number is 7 (that compares to the 5th Bell number, which from the Stirling Triangle is seen to be $1 + 15 + 25 + 10 + 1 = 52$). There is no simple exact formula for the nth partition number – there is a complex one, which is itself based on a beautiful approximation due to the Indian genius Srinivasa Ramanujan (1887–1920).

One simple symmetry regarding partitions is that the number of partitions of n into m parts is equal to the number of partitions of n in which the largest part is m. One way of seeing that this is true is through the *Ferrar's graph* (or *Young diagram*) of the partition, which is no more than the representation of the partition as a corresponding array of dots in which the rows are listed by decreasing size.

In the example shown in Figure 6 we have represented 17 partitioned as $5 + 4 + 4 + 2 + 1 + 1$. Note how the columns are also listed in decreasing order from left to right. If we reflect the array along the diagonal running from top left to bottom right, we recover a second Ferrar's graph as shown, which can be interpreted as the partition $17 = 6 + 4 + 3 + 3 + 1$. A similar reflection of the second graph returns you to the first and we say that the two corresponding partitions are *dual* to one another. This symmetry allows us to see that the numbers of partitions of two corresponding types are equal: the dual of a partition in which m, say, is the largest number (so the top row has m dots) is a partition with m rows, which corresponds to a partition into m numbers. For example, the number of partitions of 17 into 6 numbers therefore equals the number of partitions of 17 in which 6 is the largest number that occurs.

Mathematicians always have an eye out for these kind of symmetries that often arise in enumeration problems. Another example of this type occurs in relation to the Bertrand–Whitworth

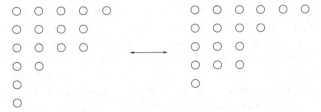

6. Dual partitions of 17 = $5 + 4 + 4 + 2 + 1 + 1 = 6 + 4 + 3 + 3 + 1$

ballot problem where the votes for two electoral candidates are counted with the winner taking p votes and the loser q votes, say. A clever geometric argument using what is known as the *Reflection Principle* shows that the proportion of counts where the winner leads the counting throughout the night equals the winner's final margin of victory divided by the total number of votes cast: $\frac{p-q}{p+q}$. This in turn is equal to the proportion of counts where the eventual winning margin is never attained until the very last vote is counted. The reason why these two numbers must be equal is that the two types of count are dual to one another in that the reversal of the order of votes in a count of the first type gives a count of the second type, and vice versa.

Hailstone numbers

Although not a counting tool, the hailstone numbers are intriguing as they are also defined recursively but have more of a flavour of the aliquot sequences that we met in Chapter 3. The following question goes by several names, the *Collatz Algorithm*, the *Syracuse Problem*, or sometimes just the $3n + 1$-*problem*, and it is simply the observation that, beginning with any number n, the following process always seems to end with the number 1. If n is even, divide it by 2, while if n is odd, replace it by $3n + 1$. For example, beginning with $n = 7$ we are led by the rules through the following sequence:

$$7 \to 22 \to 11 \to 34 \to 17 \to 52 \to 26 \to 13 \to 40$$
$$\to 20 \to 10 \to 5 \to 16 \to 8 \to 4 \to 2 \to 1$$

And so the conjecture is true for $n = 7$, and indeed it has been verified for all n up beyond a million million. Things are different if you fiddle with the rules: for instance, replacing $3n + 1$ by $3n - 1$ results in a cycle:

$$7 \to 20 \to 10 \to 5 \to 14 \to 7 \to \cdots.$$

The sequences of numbers that arise from these calculations behave like hailstones in that they rise and fall erratically over a long period but eventually, it seems, always hit the ground. Of the first 1000 integers, more than 350 have a hailstone maximum height of 9232 before collapsing to 1. This will happen once you run into a power of 2, for they are exactly the numbers that cause you to fall straight down to ground level without encountering any more updrafts.

All sorts of intriguing features can be discerned in graphs and plots based on the hailstone sequences reminiscent of other chaotic patterns that arise in maths and physics. Typing 'hailstone numbers' into your favourite search engine will provide you with a wealth of information, often intriguing, sometimes speculative, but generally inconclusive.

Chapter 6
Below the waterline of the number iceberg

Introduction

The counting numbers, 1, 2, 3, $\cdots$ are just the tip of the number iceberg. This tip is of course the first part we discover, and for a time we might believe there is no more to the iceberg than the tip, especially if we remain reluctant to look below the waterline. In the course of this chapter, we first introduce the negative integers and coupling this extension with fractions, both positive and negative, gives the collection we call the set of *rational numbers*. This number collection is often pictured to lie along the *number line*, with the positive numbers lying to the right of zero, with their negatives forming the mirror image to the left. However, the number line turns out to be the home of other numbers that cannot be expressed as fractions, such as $\sqrt{2}$ and π, to take two examples. The set of *real numbers* is the name for the collection of all numbers on the number line, which are those that can be represented by decimal expansions of any kind, as shown in Figure 7.

However, one of the great achievements of the 19th century was the full realization that the true domain of number is not one-, but rather is two-dimensional. The plane of the complex numbers is the natural arena of discourse for much of mathematics. This has been brought home to mathematicians and scientists through

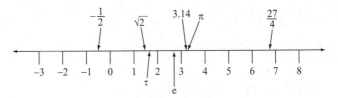

7. Central portion of the number line near 0

problem solving – to be able to carry out the investigations required to solve real-world problems, many of which seem to be only about ordinary counting numbers, it becomes necessary to expand your number horizon. The explanation as to how this extra dimension emerges will come towards the end of this chapter and be explored further in Chapter 8.

Pluses and minuses

The *integers* is the name applied to the set of all whole numbers, positive negative, and zero. This set, often symbolized by the letter **Z**, is therefore infinite in both directions:

$$\{\cdots -4, -3, -2, -1, 0, 1, 2, 3, 4, \cdots\}.$$

The integers are often pictured as lying at equally spaced points along a horizontal number line, in the order indicated. The additional rules that we need to know in order to do arithmetic with the integers can be summarized as follows:

(a) to add or subtract a negative integer, $-m$, we move m spaces to the left in the case of addition, and m spaces to the right for subtraction;

(b) to multiply an integer by $-m$, we multiply the integer by m, and then change sign.

In other words, the direction of addition and subtraction of negative numbers is the opposite to that of positive numbers, while multiplying a number by −1 swaps its sign for the alternative. For example, $8 + (−11) = −3, 3 \times (−8) = −24$, and $(−1) \times (−1) = 1$.

You should not be troubled by this last sum. First, it is reasonable that multiplying a negative number by a positive one yields a negative answer: when a debt (a negative amount) is subject to interest (a positive multiplier greater than 1) the outcome is greater debt, that is to say a larger negative number. We are all well aware of this. That multiplication of a negative number by another negative number should have the opposite outcome, that is a positive result, would then appear consistent. The fact that the product of two negative numbers is positive can readily be given formal proof. The proof is based on the assumptions that we want our expanded number system of the integers to subsume the original one of the natural numbers, and that the augmented system should continue to obey all the normal rules of algebra. Indeed, the result on the product of two negatives follows from the fact that any number multiplied by zero equals zero. (This too is not an assumption but rather is also a consequence of the laws of algebra.) For we now have:

$$−1 \times (−1 + 1) = −1 \times 0 = 0;$$

if we then multiply out the brackets, we see that in order that the left-hand side equal zero, $(−1) \times (−1)$ must take the opposite sign to $(−1) \times 1 = −1$; in other words $(−1) \times (−1) = 1$.

Fractions and rationals

In a similar way that subtraction leads to the negative numbers, the operation of division also leads us out of the set of natural counting numbers into the larger realm of fractions. However, the nature of the new arithmetic we encounter is of a different character. When adding or subtracting, fractions with different denominators (bottom lines) are incompatible. The fractions in question need to be expressed with a common denominator before the sum can be completed. Multiplication is a comparatively simple process in that we only need to multiply the numerators (top lines) and denominators together in order to get the answer. Division is the inverse operation to multiplication so that division by n corresponds to multiplication by the reciprocal, $\frac{1}{n}$. In general, this carries over to fractions in that to divide by the fraction $\frac{m}{n}$ we multiply by its reciprocal, $\frac{n}{m}$, for that reverses the effect of multiplication by $\frac{m}{n}$.

The Ancient Egyptians were only happy with *unit fractions*, which are those that are simple reciprocals of whole numbers, $\frac{1}{2}$, $\frac{1}{3}$, $\frac{1}{4}$ etc. (although they retained a special symbol for $\frac{2}{3}$). A fraction such as $\frac{3}{4}$ was not thought of as a meaningful entity in its own right, and they would record this quantity as the sum of two reciprocals: $\frac{3}{4} = \frac{1}{2} + \frac{1}{4}$. (The notation for fractions used here is of course the modern European type, which has its origins in Greek mathematics.) It is not, however, obvious that it is necessarily possible to write any fraction as the sum of a number of *different* unit fractions, which is what they insisted on. It can, however, always be done and explaining this will allow you to brush up your skills on dealing with fractions.

If you wish to find an Egyptian decomposition of a fraction such as $\frac{9}{20}$, you need only subtract the largest unit fraction you can from the given number, and repeat this process until the remainder is itself a unit fraction. This will always work, and the number of fractions involved never exceeds the numerator of your original fraction. This is because, at each stage, the numerator of the fraction that still remains is always less than the previous one: not obvious but true. In this example the first stage will give:

$$\frac{9}{20} - \frac{1}{3} = \frac{27}{60} - \frac{20}{60} = \frac{7}{60};$$

next we find that the largest unit fraction less than $\frac{7}{60}$ is $\frac{1}{9}$. (To test this, compare the result of the cross-multiplication: $\frac{1}{9} < \frac{7}{60}$ because $1 \times 60 = 60 < 63 = 7 \times 9$.) Subtracting again, we see that

$$\frac{9}{20} - \frac{1}{3} - \frac{1}{9} = \frac{7}{60} - \frac{1}{9} = \frac{21}{180} - \frac{20}{180} = \frac{1}{180}$$

and so we recover the Egyptian decomposition:

$$\frac{9}{20} = \frac{1}{3} + \frac{1}{9} + \frac{1}{180}.$$

This greedy approach of always subtracting the largest unit fraction available does work but may not yield the shortest decomposition there is, as we can see even in this case as $\frac{9}{20} = \frac{1}{4} + \frac{1}{5}$. This two-fraction decomposition of $\frac{9}{20}$ can be found, however, through use of the technique of the Akhmim papyrus, a Greek parchment discovered at the city of Akhmim on the Nile and dated to AD 500–800. In modern notation, the trick can be expressed as the readily verified algebraic identity:

$$\frac{m}{pq} = \frac{m}{p(p+q)} + \frac{m}{q(p+q)}.$$

Applying this with $m = 9$, $p = 4$, $q = 5$ immediately gives us $\frac{9}{20} = \frac{9}{4 \times 9} + \frac{9}{5 \times 9} = \frac{1}{4} + \frac{1}{5}$.

It may well have been years since you last added even simple fractions together because nearly all practical arithmetic is now carried out in decimal format. The use of decimal fractions is found in ancient China and medieval Arabic nations but only came into widespread use in Europe in the latter part of the 16th century when serious efforts were made to improve practical methods of computation. There is a price to be paid, however, for this commitment to decimal forms. In normal base ten arithmetic we exploit the fact that any number can be written as a sum of multiples of powers of ten. When expressing a fraction as a decimal, we are attempting to write the number as a sum of powers of $\frac{1}{10} = 0.1$. Unfortunately, even for very simple fractions such as $\frac{1}{3}$, this cannot be done, and the decimal expansion goes on without end: $\frac{1}{3} = 0.333\ldots$. In practice, we appreciate that by truncating the decimal expansion after a certain number of places (depending on the accuracy we demand), we can get by with the resulting *terminating decimal* that approximates the exact fraction. Any inaccuracy is trivial in comparison with the convenience of carrying out all our number work in the standard base ten frame of reference. Decimal expansions can be thought of as the closest we can get to having a single common denominator for all fractions.

It is natural to ask, though, which fractions will have terminating expansions (and which will not)? The answer is, not very many. More often than not, the decimal expansion of a fraction goes into a recurring pattern: $\frac{3}{22}$ = 0.1363636... with the 36 part repeating forever. Every fraction generates a recurring decimal in one way or another, although in the case of a terminating decimal such as $\frac{1}{2}$ = 0.5, the recurring part is simply an unending string of zeros: $\frac{1}{2}$ = 0.5000···, and so is not explicitly mentioned. In any event, the length of the recurring block in a recurring decimal expansion is no longer than one less than the value of the denominator. This can be seen by considering what happens when we carry out the corresponding long division sum: if the denominator is n, then the remainder after each step in the division takes on one of the values $0, 1, \cdots, n-1$. If at some stage the remainder is 0, the division terminates and so does the decimal expansion: for example, $\frac{11}{40}$ is exactly equal to 0.275. Otherwise the division continues forever, but once a remainder is repeated, which is inevitable, we shall be forced into the same cycle of divisions once more, thus giving a recurring pattern whose block can be no longer than $n-1$. The expansion will terminate exactly when the denominator is a product of the prime factors 2 and 5 of our base 10 but not if there is any other factor involved. For example, fractions with denominators 16, 40, and 50 are terminating, but fractions like $\frac{1}{14}$ and $\frac{1}{15}$ will not terminate because the respective prime factors of 7 and 3 in their denominators force the expansion into a recurring cycle.

This does show, however, that whether or not a fraction's expansion terminates is not determined just by the number itself but rather depends on the relationship of the number to the base in which you are operating. If, for instance, we worked in *ternary* (base three) then 0.1 would represent $\frac{1}{3}$, as the 1 after the decimal point would stand for $\frac{1}{3}$, and not $\frac{1}{10}$, the way it does in decimal expansions.

The reverse process of turning a recurring decimal back into a fraction is also quite simple, showing that there is a one-to-one correspondence between fractions and recurring decimals, and we can use whichever representation best suits our current purpose. A simple example is as follows: let $a = 0.212121\cdots$. Since the length of the recurring block is 2, we can simplify this, as you will see, by multiplying by $10^2 = 100$ to obtain $100a = 21.212121\cdots$. This has been set up so that, upon subtraction, the recurring parts of the two numbers a and $100a$, being identical, will cancel, allowing us to infer that $99a = 21$, whence $a = \frac{21}{99} = \frac{7}{33}$.

This kind of trick is often used to simplify an expression that involves an infinite repeating process. For example, consider the following little monster:

$$a = \sqrt{2\sqrt{5\sqrt{2\sqrt{5}}}}\cdots$$

By squaring, and then squaring again, the left-hand side becomes a^4, while the expression on the right gives:

$$a^4 = 2^2 \times 5 \times \sqrt{2\sqrt{5\sqrt{2\sqrt{5}}}}\cdots$$

Since what follows the 5 is another copy of the expression for a, we infer that $a^4 = 20a$ so that $a^3 = 20$ or, if you prefer, a is the cube root of 20. We will call on this technique again in Chapter 7 when we introduce so-called continued fractions.

Does the class of fractions provide us with all the numbers we could ever need? As mentioned earlier, the collection of all fractions, together with their negatives, form the set of numbers known as the rationals, that is all numbers that result from whole numbers and the ratios between them. They are adequate for arithmetic in that any sum involving the four basic arithmetic operations of addition, subtraction, multiplication, and division will never take you outside the world of rational numbers. If we are happy with that, this set of numbers is all we require. However, we explain in the next section how numbers such as *a* above are not rational.

Irrationals

The word *irrational* applied to a number *a* means simply that the number is not rational, that is to say *cannot* be written as a fraction. Irrational numbers were first discovered a very long time ago, in ancient Greece. Pythagoras understood the irrational nature of $\sqrt{2}$. The Greeks did not think in terms of decimal expansions but were happy to recognize a length constructed in the geometry of straightedge and compass as representing a real quantity. In particular, Pythagoras' Theorem tells us that the longer side of a right-angled triangle whose shorter sides are each of length 1 unit is exactly equal to the square root of 2.

Pythagoras was able to prove that the square root of 2 was not equal to any fraction, thereby showing that irrational numbers truly exist. In particular, you cannot exactly measure the diagonal of a square with the same units with which you measure the side, for if you could the diagonal would be an exact fractional multiple of the side, in which case $\sqrt{2}$ would be equal to this fraction. The two lengths are however fundamentally incompatible, or *incommensurable* as they are described in the classical texts. The story is the same for π, which is approximately equal to the fraction $\frac{22}{7}$, but is different from it, and from any fraction that you care to nominate. (However, the easily remembered 'double 1, double 3, double 5' ratio: $355/113 = 3.1415929\cdots$ accurately approximates the value of π to better than one part in one million.)

Although it is very difficult to prove that π is irrational, the question for the square root of 2 can be settled easily by a simple contradiction argument. First, we note that for any number c, the highest power of 2 that is a factor of c^2, is twice the highest power of 2 that is a factor of c, and so in particular the highest power of 2 that divides any square must itself be an even number. For example, $24 = 2^3 \times 3$ while $576 = 24^2 = 2^6 \times 3^2$, and in this case the highest power of 2 dividing the number does indeed double from 3 to 6 when we take the square. This is always the case, and indeed applies not only to powers of 2 but to any prime factor of the original number.

Suppose now that $\sqrt{2}$ were equal to the fraction $\frac{a}{b}$. Squaring both sides of this equation allows us to deduce that $2 = \frac{a^2}{b^2}$, which gives $2b^2 = a^2$. By the previous observation, the highest power of 2 that divides the right-hand side of this equation is even, while the highest power that divides the left-hand side is odd (because of the presence of the extra 2). This shows the equation to be nonsense, and so it must not be possible to write $\sqrt{2}$ as a fraction in the first place. Like Pythagoras, we come face to face with the irrationals.

Arguments along these lines allow us to show that quite generally, when we take the square root (or indeed the cube or a higher root) of a number, the answer, if not a whole number, is always irrational, thus explaining why the decimal displays on your calculator never show a recurring pattern when asked to calculate such a root.

Pythagoras discovered that in order to do his mathematics, he required a wider field of numbers than mere fractions. The Greeks regarded a number to be 'real' if its length could be constructed from a standard unit interval using only a straightedge (not a marked ruler, just an edge) and compass. It turns out that although the square root operation does introduce irrationals, the full collection itself does not go very far beyond the rational. The set of euclidean numbers, as we shall refer to them, are all those that can be arrived at from the number 1 through carrying out any or all of the four operations of arithmetic and the taking of square roots, any number of times. For example, the number $\sqrt{7 - \sqrt{4/3}}$ is therefore a number of this kind. Even cube roots are beyond the grasp of the euclidean tools. This was the basis of perhaps the first great unsolved problem in mathematics. The first of the three Delian Problems as they were known was the call to construct the cube root of 2, using only straightedge and compass. Legend has it that this was the task set by the god when the citizens of Delos consulted the Oracle of Delphi to learn what they should do in order to banish the plague from Athens – the problem was put in the form of exactly doubling the volume of an altar that was a perfect cube.

This problem remained untouchable in classical times. That the cube root of 2 lies outside the range of the euclidean tools was only settled in 1837 by Pierre Wantzel (1814–38), as it requires a precise algebraic description of what is possible using the classical tools in order to see that the cube root of 2 is a number of a fundamentally different type. It does indeed come down to showing that you can never manufacture a cube root out of square roots and rationals. When put that way, the impossibility sounds more plausible. However, that in no way constitutes a proof.

Transcendentals

Within the class of irrationals lies the mysterious family of transcendental numbers. These numbers do *not* arise through the ordinary calculations of arithmetic and the extraction of roots. For the precise definition, we first introduce the complementary collection of *algebraic numbers*, which are those that solve some polynomial equation with integer coefficients: for example $x^5 - 3x + 1 = 0$ is such an equation. The *transcendentals* are then defined to be the class of non-algebraic numbers.

It is not at all clear that there are any such numbers. However, they do exist and they form a very secretive society, with those in it not readily divulging their membership of the club. For example, the number π is an instance of a transcendental but this is not a fact that it openly reveals. It will be explained in the next chapter when we explore the nature of infinite sets why it is that 'most' numbers are transcendental, in a sense that will be made precise.

For the time being, I will settle for introducing perhaps the most famous transcendental of all, the number $e = 2.71828\ldots$. This number arises constantly in higher mathematics and calculus: it is the base of the so-called natural logarithm, the function that tells you the area under the graph of the reciprocal function. It is also the limiting value of the sequence of numbers you get when you raise the ratio of two consecutive integers, $\frac{n+1}{n}$ ($= 1 + \frac{1}{n}$), to the power n. (Ask your calculator for the value of $(129/128)^{128}$ – you can 'fast exponentiate' this, just calculate $129/128$ and then square 7 times, as $2^7 = 128$.)

This sequence arises when we consider the problem of the limiting value of a compound interest rate as you reduce the interval of repayment shorter and shorter from annually, to monthly, to daily, and so on. To best illustrate the point, suppose that interest is paid at an annual rate of 100%, compounding in n instalments per year, which means that your initial investment is multiplied by the factor $(1 + \frac{1}{n})$, n times in all, throughout the course of the year. Your principal will then be multiplied by the factor $(1 + \frac{1}{n})^n$. The more often you are paid interest, the more you will earn as you begin to collect interest on your interest earlier and earlier as n becomes higher and higher. However, as n increases, the effective APR (Annual Percentage Rate) does not increase beyond all bounds but rather approaches a ceiling, an upper limit as mathematicians call it. This limiting multiplier that would apply to your principal in the continuous interest case is the limiting value, as n increases, of the number

$$\left(1 + \frac{1}{n}\right)^n \to e = 2.71828\ldots.$$

Another way in which the mysterious e arises is through the sum of the reciprocals of the factorials, and this gives a way of calculating e to a high degree of accuracy as this series converges rapidly because its terms approach zero very quickly indeed:

$$e = 1 + \frac{1}{1!} + \frac{1}{2!} + \frac{1}{3!} + \frac{1}{4!} + \cdots$$

This representation allows you to show by a relatively simple contradiction argument, outlined here, that e is an irrational number. We suppose that the preceding series for e equals a fraction $\frac{p}{q}$ and then we multiply both sides by $q!$. The left-hand side is then an integer but the right-hand side consists of terms that are integers followed by an infinite sequence of non-integral terms. By comparing to a simple geometric series, we deduce that this 'tail' sums to less than 1, and so the right-hand side cannot be a whole number, and therein lies the required contradiction. Showing that e is not just irrational, but transcendental, requires quite a bit more work.

The relationship of e with the factorials also manifests itself in a remarkable formula of the Scottish mathematician James Stirling (1692–1770), after whom Stirling numbers (see Chapter 5) are named. He showed that as n increases, the value of $n!$ is approximated better and better by the expression $\sqrt{2\pi n}(\frac{n}{e})^n$.

Since e crops up in a variety of distinct and fairly simple ways, it persistently appears throughout mathematics, often where you would not expect to meet it. For example, take two well-shuffled packs of playing cards, turn over the top card of each deck, and compare. Continue doing this until you have exhausted the packs. What are the chances that, at some stage, there is a perfect match? That is to say, on one turn or another the cards showing are exactly the same, be they the seven of clubs, queen of hearts, or whatever. It works out that the proportion of times this experiment yields at least one such match is as near as makes no difference to $\frac{1}{e}$, which is about 36.8%. This comes about through application of what is known as the *inclusion–exclusion principle*, which arrives at the solution through a sum of terms each of which represent alternating corrections and reverse corrections. In this example, the principle furnishes the series of reciprocals of the factorials but this time with alternating signs, which converges to $\frac{1}{e}$.

The real and the imaginary

The first five chapters of this *Very Short Introduction* dealt mainly with positive integers. We emphasized factorization properties of integers, which led us to consider numbers that have no proper factorizations, which are the primes, a set that occupies a pivotal position in modern cryptography. We also looked at particular types of numbers, such as the Mersenne primes, which are intimately connected with perfect numbers and took time to introduce some special classes of integers that are important in counting certain naturally occurring collections. Throughout all this, the backdrop was the system of integers, which are the counting numbers, positive, negative, and zero.

In this chapter we have gone beyond integers, first to the rationals (the fractions, positive and otherwise), then to the irrationals, and within the class of irrationals we have identified the transcendental numbers. The underlying system in which all this is taking place is the system of the *real numbers*, which can be thought of as the collection of all possible decimal expansions. Any positive real number can be represented in the form $r = n.a_1 a_2 \cdots$, where n is a non-negative integer and the decimal point is followed by an infinite trail of digits. If this trail eventually falls into a recurring pattern, then r is in fact rational and we have shown how to convert this representation into an ordinary fraction. If not, then r is irrational, so the real numbers come in those two distinct flavours, the rational and the irrational.

In our mathematical imaginations, we often picture the real numbers as corresponding to all the points along the number line as we look out from zero, to the right for the positive reals, and to the left for the negative reals. This leaves us with a symmetrical picture with the negative real numbers being a mirror image of the positive reals, and this symmetry is preserved when dealing with addition and subtraction – but not with multiplication. Once we pass to multiplication, the positive and negative numbers no longer have equal status as the number 1 is endowed with a property that no other number possesses, for it is the *multiplicative identity*, meaning that $1 \times r = r \times 1 = r$ for any real number r. Multiplication by 1 fixes the position of any number, but in contrast multiplication by -1 swaps a number for its mirror image on the far side of 0. Once multiplication enters the scene, the fundamental differences in the nature of positive and negative numbers are revealed. In particular, negative numbers lack square roots within the real number system because the square of any real number is always greater than or equal to zero.

This is the cue for imaginary numbers to make their entrance. This topic is one that we shall take up again in the final chapter; for the time being, we will just make some introductory comments.

These numbers arise through the search for solutions to simple polynomial equations. In particular, since the square of any real number is never negative, we can find no solution to the equation $x^2 = -1$. Undaunted, mathematicians invented one, denoted by i, which is endowed with that property, so that $i^2 = -1$. At first sight, this seems artificial and arbitrary but it is not too much different from the kind of behaviour we have indulged in before. After all, while recognizing that the counting numbers $1, 2, 3, \cdots$ are pre-eminent, in order to deal smoothly with general number matters we are led to the wider number system of the rationals, which is the collection of all fractions, positive, negative, and zero. We then find, however, that we have no solution to the equation $x^2 = 2$, as we have shown that the square of a rational number cannot exactly equal 2. To deal with this, we have to 'invent' $\sqrt{2}$. At this point we could take the alternative attitude and say that we have proved that the square root of 2 simply does not exist and that is the end of the matter. However, few would feel happy to haul up the drawbridge in this way. The ancient Greeks certainly were not content to let things stand at that, for they could construct a length representing $\sqrt{2}$ with compass and straightedge and so the number was, to their way of thinking, definitely real and any mathematical system that denied this was inadequate.

We might agree with Pythagoras for quite a different reason. We may react by saying that we can approximate $\sqrt{2}$ to any degree of accuracy via its decimal expansion: $\sqrt{2} = 1.414213\cdots$, and so $\sqrt{2}$ is the number that is represented exactly by the totality of this expansion. A modern person might find more force in this argument and so insist for this reason that the number system needs to be expanded beyond the rationals.

However, at first glance we might say that things are different when it comes to $\sqrt{-1}$ as there seems no immediate need to worry about its non-appearance among the collection of numbers that we habitually call 'real'. It transpires though that as our mathematics progresses a little further, the need for imaginary numbers becomes very pressing, and any initial reluctance to deal with them is dispelled as our understanding of things mathematical grows.

This first struck home in the 16th century when Italian mathematicians learnt how to solve cubic and fourth-degree polynomial equations in a fashion that extended that used to solve quadratic equations. The Cardano method, as it came to be known, would often involve square roots of negatives even though the solutions to the equations eventually turned out to be positive integers. By stages from this point, the use of *complex numbers*, which are those of the form $a + bi$, where a and b are ordinary real numbers, was shown to facilitate a variety of mathematical calculations. For example, in the 18th century Euler revealed and exploited the stunning little equation $e^{i\pi} = -1$, which cannot fail to surprise anyone on their first encounter.

Around the beginning of the 19th century, the geometric interpretation of complex numbers as points in the coordinate plane (the standard system of xy-coordinates), was investigated by Wessell and Argand, from which point the use of the 'imaginary' became accepted as normal mathematics. Identifying the complex number $x + iy$ with the point with coordinates (x, y) allows examination of the behaviour of complex numbers in terms of the behaviour of points in the plane, and this proves to be very illuminating. The theory of so-called *complex variables*, whose subject matter is represented by functions of complex numbers, rather than just real numbers, flourished spectacularly in the hands of Augustin Cauchy (1789–1857). It is now a cornerstone of mathematics, underpins much of electrical signal theory, and the entire field of X-ray diffraction is built on complex numbers. These numbers have proved to have real meaning, and moreover the system is complete in that *every* polynomial equation has its full complement of solutions within the system of complex numbers. We shall return to these matters in the final chapter. Before doing that, however, we shall in the next chapter look more closely at the infinite nature of the real number line.

Chapter 7
To infinity and beyond!

Infinity within infinity

It was the great 16th-century Italian polymath Galileo Galilei (1564–1642) who was first to alert us to the fact that the nature of infinite collections is fundamentally different from finite ones. As alluded to on the first page of this book, the size of a *finite* set is smaller than that of a second set if the members of the first can be paired off with those of just a portion of the second. However, infinite sets by contrast can be made to correspond in this way to subsets of themselves (where by the term *subset* I mean a set within the set itself). We need go no further than the sequence of natural counting numbers $1, 2, 3, 4, \cdots$ in order to see this. It is easy to describe any number of subsets of this collection that themselves form an infinite list, and so are in a one-to-one correspondence with the full set (see Figure 8): the odd numbers, $1, 3, 5, 7, \cdots$, the square numbers, $1, 4, 9, 16, \cdots$ and, less obviously, the prime numbers, $2, 3, 5, 7, \cdots$, and in each of these cases the respective complementary sets of the even numbers, the non-squares, and the composite numbers are also infinite.

The Hilbert Hotel

This rather extraordinary hotel, which is always associated with David Hilbert (1862–1943), the leading German mathematician of his day, serves to bring to life the strange nature of the infinite. Its

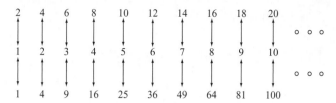

8. The evens and the squares paired with the natural numbers

chief feature is that it has infinitely many rooms, numbered
$1, 2, 3, \cdots$, and boasts that there is always room at Hilbert's Hotel.

One night, however, it is in fact full, which is to say each and every
room is occupied by a guest and much to the dismay of the desk
clerk, one more customer fronts up demanding a room. An ugly
scene is avoided when the manager intervenes and takes the clerk
aside to explain how to deal with the situation: tell the occupant of
Room 1 to move to Room 2 says he, that of Room 2 to move into
Room 3, and so on. That is to say, we issue a global request that
the customer in Room n should shift into Room $n + 1$, and this will
leave Room 1 empty for this gentleman!

And so you see, there *is* always room at the Hilbert Hotel. But how
much room?

The next evening, the clerk is confronted with a similar but more
testing situation. This time a spaceship with 1729 passengers
arrives, all demanding a room in the already fully occupied hotel.
The clerk has, however, learned his lesson from the previous night
and sees how to extend the idea to cope with this additional group.
He tells the person in Room 1 to go to Room 1730, that of Room 2
to shift to Room 1731, and so on, issuing the global request that
the customer in Room n should move into Room number $n + 1729$.
This leaves Rooms 1 through to 1729 free for the new arrivals, and
our clerk is rightly proud of himself for dealing with this new
version of last night's problem all by himself.

The final night, however, the clerk again faces the same situation – a full hotel, but this time, to his horror, not just a few extra customers show up but an infinite space coach with infinitely many passengers, one for each of the counting numbers 1, 2, 3. ⋯. The overwhelmed clerk tells the coach driver that the hotel is full and there is no conceivable way of dealing with them all. He might be able to squeeze in one or two more, any finite number perhaps, but surely not infinitely many more. It is plainly impossible!

An infinite riot might have ensued except again for the timely intervention of the manager who, being well versed in Galileo's lessons on infinite sets, informs the coach driver that there is no problem at all. There is always room at Hilbert's Hotel for anyone and everyone. He takes his panicking desk clerk aside for another lesson. All we need do is this, he says. We tell the occupant of Room 1 to shift into Room 2, that in Room 2 to shift to *Room 4*, that in Room 3 to go to *Room 6*, and so on. In general the global instruction is that the occupant of Room n should move into Room $2n$. This will leave all the odd numbered rooms empty for the passengers of the infinite space coach. No problem at all!

The manager seems to have it all under control. However, even he would be caught out if a spaceship turned up that somehow had the technology to have one passenger for each point in the continuum of the real line. One person for every decimal number would totally overrun Hilbert's Hotel, and we shall see why in the next section.

Cantor's comparisons

All this may be surprising the first time you think about it, but it is not difficult to accept that the behaviour of infinite sets may differ in some respects from finite ones, and this property of having the same size as one of its subsets is therefore a case in point. In the 19th century, however, Georg Cantor (1845–1918) went much further and discovered that not all infinite sets can be regarded as having equally many members. This revelation was unexpected in nature but is not hard to appreciate once your attention is drawn to it.

Cantor asks us to think about the following. Suppose we have any infinite list L of numbers $a_1, a_2, \cdots$ thought of as being given in decimal form. Then it is possible to write down another number, a, that does not appear anywhere in the list L: we simply take a to be different from a_1 in the first place after the decimal point, different from a_2 in the second decimal place, different from a_3 in the third decimal place, and so on – in this way, we may build our number a making sure it is not equal to any number in the list. This observation looks innocuous but it has the immediate consequence that it is *absolutely impossible* for the list L to contain *all* numbers, because the number a will be missing from L. It follows that the set of all *real numbers*, that is all decimal expansions, cannot be written in a list, or in other words *cannot* be put into a one-to-one correspondence with the natural counting numbers, the way we saw in Figure 8. This line of reasoning is known as *Cantor's Diagonal Argument*, as the number a that lies outside the set L is constructed by imagining a list of the decimal displays of L as in Figure 9 and defining a in terms of the diagonal of the array.

There is some subtlety here, for we might suggest that we can easily get around this difficulty by simply placing the missing number a at the front of L. This creates an enlarged listing M containing the annoying number a. However, the underlying problem has not gone away. We can apply Cantor's construction again to introduce a fresh number b that is not present in the new list M. We can of course continue to augment the current list as before any number of times, but Cantor's point remains valid: although we can keep creating lists that contain additional numbers that were previously overlooked, there can never be one specific list that contains every real number.

$$a_1 = 0.\,a_{11}\;a_{12}\;a_{13}\;a_{14}\;\circ\;\circ\;\circ\;a_{1k}\;\circ\;\circ\;\circ$$

$$a_2 = 0.\,a_{21}\;a_{22}\;a_{23}\;a_{24}\;\circ\;\circ\;\circ\;a_{2k}\;\circ\;\circ\;\circ$$

$$a_3 = 0.\,a_{31}\;a_{32}\;a_{33}\;a_{34}\;\circ\;\circ\;\circ\;a_{3k}\;\circ\;\circ\;\circ$$

$$a_k = 0.\,a_{k1}\;a_{k2}\;a_{k3}\;a_{k4}\;\circ\;\circ\;\circ\;a_{kk}\;\circ\;\circ\;\circ$$

9. Cantor's number a differs from each a_k in the kth decimal place

The collection of all real numbers is therefore larger in a sense than the collection of all positive integers. Even though both are infinite, the sets cannot be paired off together the way the even numbers can be paired with the list of all counting numbers. Indeed, if we apply Cantor's Diagonal Argument to a putative list of all numbers in the interval 0 to 1, the missing number a will also lie in this range. Therefore, we likewise conclude that this collection will also defy every attempt at listing it in full. I mention this as we shall make use of that fact shortly.

Cantor's result is rendered all the more striking by the fact that many other sets of numbers can be put into an infinite list, including the Greeks' euclidean numbers. A little ingenuity is involved, but once a couple of tricks are learned, it is not hard to show that many sets of numbers are *countable*, which is the term we use to mean the set can be listed in the same fashion as the counting numbers. Otherwise an infinite set is called *uncountable*.

For example, let us take the set of all integers **Z**, which comes to us naturally as a kind of doubly infinite list. We can, however, rearrange it into a row with a starting point: 0, 1, −1, 2, −2, 3, −3, ⋯ by pairing each positive integer with its opposite, we create a list where every integer appears – none will escape. More surprisingly, we can also do the same with the rationals: start with 0, then list all the rationals that can be written using all numbers no more than 1 (which are 1 and −1), then those that involve no number higher than 2 (which are 2, −2, $\frac{1}{2}$, −$\frac{1}{2}$), then those that only use numbers up to 3, and so on. In this way, the fractions (positive, negative, and zero) can be arranged in a sequence in which they are all present and accounted for. The rationals therefore also form a countable set, as do the euclidean numbers, and indeed if we consider the set of all numbers that arise from the rationals through taking roots of any order, the collection produced is still countable. We can even go beyond this: the collection of all algebraic numbers (first mentioned in Chapter 6), which are those that are solutions of ordinary polynomial equations, form a collection that can, in principle, be arrayed in an infinite list: that is to say, it is possible to describe a systematic listing that sweeps them all out. (The proof is along the lines of the argument that works for the rationals.)

What we have allowed to happen in casually accepting any decimal expansion is to open the door to what are known as the transcendental numbers, those numbers that lie beyond those that arise through euclidean geometry and ordinary algebraic equations. Cantor's argument shows us that transcendental numbers exist and, in addition, there must be infinitely many of them, for if they formed only a finite collection, they could be placed in front of our list of algebraic numbers (the non-transcendentals), so yielding a listing of all real numbers, which we now know is impossible. What is striking is that we have discovered the existence of these strange numbers without identifying a single one of them! Their existence was revealed simply through comparing certain infinite collections to one another. The transcendentals are the numbers that fill the huge void between the more familiar algebraic numbers and the collection of all decimal expansions: to borrow an astronomical metaphor, the transcendentals are the dark matter of the number world.

In passing from the rationals to the reals, we are moving from one set to another of *higher cardinality* as mathematicians put it. Two sets have the same cardinality if their members can be paired off, one against the other. What can be shown using the Cantor argument is that any set has a smaller cardinality than the set formed by taking all of its subsets. This is obvious for finite collections: indeed, in Chapter 5 it was explained that if we have a set of n members then there are 2^n subsets that can be formed in this way. But how large is the set S of all subsets of the infinite set of natural numbers, $\{1, 2, 3, \cdots\}$? This question is not only interesting in itself but also in the manner in which we arrive at the answer, which is that S is indeed uncountable.

Russell's Paradox

Suppose to the contrary that S was itself countable, in which case the subsets of the counting numbers could be listed in some order $A_1, A_2, \cdots$. Now an arbitrary number n may or may not be a member of A_n – let us consider the set A that consists of all numbers n such that n is *not* a member of the set A_n. Now A is a subset of the counting numbers (possibly the empty subset) and so appears in the aforesaid list at some point, let $A = A_j$ say. An unanswerable question now arises: is j a member of A_j? If the answer were 'yes' then, by the very way A is defined, we conclude that j is *not* a member of A, but $A = A_j$, so that is self-contradictory. The alternative is no, j is not a member of A_j, in which case, again by the definition of A, we infer that j is a member of $A = A_j$, and once more we have contradiction. Since contradiction is unavoidable, our original assumption that the subsets of the counting numbers could be listed in a countable fashion must be false. Indeed, this argument works to show that the set of all subsets of any countable but infinite set is uncountable.

This particular self-referential style argument was introduced by Bertrand Russell (1872–1970) in a slightly different context that led to what is known as Russell's Paradox. Russell applied it to the 'set of all sets that are not members of themselves', asking the embarrassing question whether or not that set is a member of itself. Again, 'yes' implies 'no' and 'no' implies 'yes', forcing Russell to conclude that this set cannot exist.

In the 1890s, Cantor himself discovered an implicit contradiction stemming from the idea of the 'set of all sets'. Indeed, Russell acknowledged that the argument of his paradox was inspired by the work of Cantor. The upshot of all this, however, is that we cannot simply imagine that mathematical sets can be introduced in any manner whatsoever, but some restrictions must be placed on how sets may be specified. Set theorists and logicians have been wrestling with the consequences of this ever since. The most satisfactory resolution of these difficulties is provided by the now standard *ZFC Set Theory* (the Zermelo-Fraenkel set theory with the *Axiom of Choice*).

The number line under the microscope

There are different ways of looking at the size of infinite sets of numbers that are revealed if we look at the distribution of the various number types that knit together to bind the number line into a continuum. The rationals may only be a countable collection of numbers, but the collection is densely packed within the line in a way that the integers plainly are not. Given any distinct numbers, a and b, there is a rational number that lies between them. The average of the two numbers, $c = \frac{a+b}{2}$, certainly is a number lying between them, but it may be irrational. However, if c is irrational we can approximate it by a rational number d, with a terminating decimal expansion, by letting d have the same decimal representation as c up to a very large number of decimal places. For example, if we take $\sqrt{2} = 1.414\ldots$, we have that $\sqrt{2}$ differs from 1.414 by less than 0.001, and each time we take another decimal place we guarantee finding a rational number that approximates $\sqrt{2}$ more accurately (on average, ten times more accurately) than the previous one. If the number of initial places in which they agree is sufficiently large, then their difference will be so small that both c and d will lie between a and b. The number of places we need to take after the decimal place will depend on just how close a and b are to each other to begin with, but it is always possible to find a rational d that does the job (see Figure 10). We say that the set of rational numbers is *dense* in the number line for just this reason. Of course we can, by the same argument, show that there is another rational, splitting the interval from a to d, say, and, in this way, we arrive at the conclusion that infinitely many rational numbers lie between any two numbers, however small the difference between these two numbers might happen to be. In particular, there is no such thing as the smallest positive fraction, for, given any positive number, there is always a rational lying between it and zero.

Not to be outdone, the set of irrationals also forms a dense set. Before explaining this, I point out that once we have identified one irrational, the Pythagorean number $\sqrt{2}$ for example, the floodgates open and we can immediately identify infinitely more. When we add a rational to an irrational the result is always an irrational. For example, $\sqrt{2} + 7$ is irrational by dint of this fact. In a similar fashion, if we multiply an irrational number by a rational number (other than 0), the result is another irrational number.

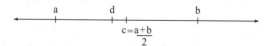

10. **Rationals separate any given positions on the number line**

(Simple contradiction arguments establish both these claims.) In particular, we can find an irrational number of size as small as we like: $t = \frac{\sqrt{2}}{n}$ is irrational for any counting number n and by taking n larger and larger we can make t as close to 0 as we please. As with the rationals, we therefore see that there is no smallest positive irrational, and hence there is no such thing as the smallest positive number.

Returning to our given numbers, a and b, once again let c be their average. If c is irrational, we have a number of the required kind (irrational). If on the other hand, c is rational, put $d = c + t$, where t is the irrational number of the previous paragraph. By what has gone before, d will also be irrational, and if we take n large enough, we can always ensure that d is so close to the average c of the two given numbers a and b that it lies between them. In this way, we see that the irrational numbers too form a dense set and, as with the rational numbers, we can infer that there are infinitely many irrational numbers lying between any two numbers on the number line.

And so the set of rationals and its complementary set of irrationals are in one way comparable (they are both dense in the number line) and in another not (the first set is countable, the second not).

Cantor's Middle Third Set

We now have a clearer idea as to how the rational and the irrational numbers interlace to form the real number line. The rational numbers form a countable set, yet are densely packed into the number line. Cantor's Middle Third Set is, by way of contrast, an uncountable subset of the unit interval that nevertheless is sparsely spread. It is the result of the following construction.

We begin with the unit interval I, that is all the real numbers from 0 up to 1 inclusive. The first step in the formation of Cantor's set is the removal of the middle third of this interval, that is all the

11. Evolution of Cantor's Middle Third Set to the 4th level

numbers between $\frac{1}{3}$ and $\frac{2}{3}$. The set that remains consists of the two intervals from 0 up to $\frac{1}{3}$ and from $\frac{2}{3}$ up to 1. At the second stage, we remove the middle third of these two intervals, at the third stage we remove the middle third of the remaining intervals, and so on. Cantor's Middle Third Set C then consists of all points of I that are *not* removed at any stage of this process.

The total length of the little intervals that remain as we pass from one stage of this process to the next is, by design, $\frac{2}{3}$ that of the previous stage; it follows that at the *nth* stage the total length of the surviving intervals is $\left(\frac{2}{3}\right)^n$. This expression approaches 0 as n increases and since the Cantor set C is the collection of all points that are left at the end of it all, it follows that the 'length', or *measure*, of C must be 0.

We might suspect that we have thrown the baby out with the bath water and that there are no points at all left in C. Is the Middle Third Set empty? The answer is a resounding no! There are infinitely many numbers left in C. This is best seen if we shift our representation of the numbers of the interval to base three 'decimals' known as *ternary*, as the whole construction is based on thirds. In base three decimals, the numbers $\frac{1}{3}$ and $\frac{2}{3}$ are respectively given by 0.1 and 0.2. By discarding the middle third of the unit interval, we have thrown away all those numbers whose ternary expansion begins with 0.1, and indeed the overall process weeds out exactly those numbers that have a 1 anywhere in their ternary expansions. The numbers in C are exactly those whose ternary expansions consist entirely of 0s and 2s. (For example, $\frac{3}{4}$ survives the infinite cull as in ternary it has the recurring expansion $0.202020\ldots$.)

Next we make an amazing observation. By taking the ternary expansion of any number c in C and replacing each instance of 2 by 1, we obtain the binary expansion of some number c' in the unit interval. This gives a one-to-one correspondence of C with the set of *all* numbers in I (written in binary). It follows that the cardinality of C is the same as that of I, and since the latter is an uncountable set (by Cantor's Diagonal Argument), it follows that the Cantor Middle Third Set is not only infinite but uncountable.

Therefore we have a set C that is in one sense negligible in size (has measure zero), but by another way of reckoning C is huge, as it has the same cardinality as I and hence of the whole real line.

What is more, far from being dense, C is *nowhere dense*. Recall that by saying that a set like the rationals is dense, we mean that whenever we take a real number a, there are rational numbers to be found in any little interval surrounding a, however small that interval might be. We say that any *neighbourhood* of a contains members of the set of rationals. The Cantor set has quite the opposite nature – numbers not in C might live their lives in the real line without ever coming across any members of C, provided they confine their experiences to a narrow enough locality around where they live. To see this, take any number a that is *not* in C, so that a has a ternary expansion that contains at least one 1: $a = 0 \cdot \ldots.1\ldots.$, with the 1 in the nth place, say. For a sufficiently tiny interval surrounding a, the numbers b in that interval have a ternary expansion that agrees with that of a up to places beyond the nth, and so *all* of them will also *not* be members of the strange set C as their ternary expansions will also contain at least one instance of 1.

On the other hand, any member a of the Cantor set will not feel too isolated, for when a looks out into any interval J that surrounds it in the number line, however small, a will find neighbours from the set C living alongside it (and numbers not in C as well). We can specify a member b of the given interval J that also lies in C by taking b to have a ternary expansion that agrees with a to a sufficiently large number of places, but with no entry being a 1. Indeed, there are uncountably many members of C in J.

In conclusion, the Cantor Middle Third Set C is as numerous as can be and, to the members of the C club, their brothers and sisters are to be seen all around them wherever they look. To the numbers not in C, however, C hardly seems to exist at all. Not one member of C is to be spotted in their exclusive neighbourhoods, and the set C itself has measure zero. To them, C is almost nothing.

Diophantine equations

Some of the principal sets of the number line may be characterized in the language of equations. The rational numbers, which form a countable set, are the numbers that arise as solutions of simple linear equations: the fraction $\frac{b}{a}$ is the solution to the equation $ax - b = 0$ (a and b are integers). Numbers like $\sqrt{2}$ that do not arise in this way are called irrational, and they form an uncountable collection that cannot be paired off with the counting numbers in the way that the rationals can. Within the set of irrationals there are the transcendentals, which are the numbers that never arise as the solutions of equations of these kinds even if we allow higher powers of x. It is known that π is an example of a transcendental number, but $\sqrt{2}$ is not, as it solves the equation $x^2 - 2 = 0$. The approach then is to define classes of numbers through the kinds of equations they solve.

An interesting line of study emerges, however, when we take the opposite tack and insist that not only the coefficients of our equations but their solutions have also to be integers. Here is a classic example.

A box contains spiders and beetles and 46 legs. How many of each kind of creature are there? This little number puzzle can be solved easily by trial, but it is instructive to note that first, it can be represented by an equation: $6b + 8s = 46$, and second that we are only interested in certain kinds of solutions to that equation, namely those where the number of beetles (b) and spiders (s) are counting numbers. In general, a system of equations is called *Diophantine* when we are restricting our solution search to special number types, typically integer or rational answers are what we are after.

There is a simple method for solving linear Diophantine equations such as this one. First, divide through the equation by the hcf of the coefficients, which are in this case 6 and 8 so their hcf is 2. Cancelling this common factor of 2 we obtain an equivalent equation, that is to say one with the same solutions: $3b + 4s = 23$. If the right-hand side were no longer an integer after performing this division, that would tell us that there were no integral solutions to the equation and we could stop right there. The next stage is to take one of the coefficients, the smaller one is normally the easiest, and work in multiples of that, in this case 3. Our equation can be written as $3b + 3s + s = (3 \times 7) + 2$; rearranging we obtain $s = (3 \times 7) - 3b - 3s + 2$. The point of this is that it shows that s has the form $3t + 2$ for some integer t. Substituting $s = 3t + 2$ into our equation and making b the subject, we get

$$3b + 4(3t + 2) = 23 \Rightarrow 3b = 15 - 12t \Rightarrow b = 5 - 4t.$$

We now have the complete solution in integers to the Diophantine equation: $b = 5 - 4t$, $s = 3t + 2$. Choosing any integral value for t will give a solution, and all solutions in integers are of this form.

Our original problem, however, was further constrained in that both b and s had to be at least zero, as negative beetles and spiders

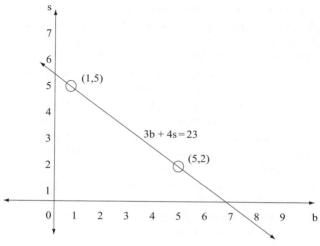

12. Lattice points on the line of a linear equation

do not exist. Hence there are only two feasible values of t, those being $t = 0$ and $t = 1$, giving us the two possible solutions of 5 beetles and 2 spiders, and 1 beetle and 5 spiders. If we interpret the puzzle as meaning that there is a plurality of both types of creature, we have the traditional solution: 5 beetles and 2 spiders.

This type of problem is called *linear* because the graph of the associated equation consists of an infinite line of points. The Diophantine problem then is to find the *lattice points* on this line, which are the points where both coordinates are integral or, if we only admit positive solutions, only lattice points in the positive quadrant will do.

However, once we allow squares and higher powers into our equations, the nature of the corresponding problems are much more varied and interesting. A classical problem of this type that has a full solution is that of finding all *Pythagorean triples*: positive integers a, b, and c such that $a^2 + b^2 = c^2$. A Pythagorean triple of course takes its name from the fact that it allows you to draw a right-angled triangle with sides of those integer lengths. The classic example is the $(3, 4, 5)$ triangle. Given any Pythagorean triple, we can generate more of them simply by multiplying all the numbers in the triple by any positive number as the Pythagorean equation will continue to hold. For example, we can double the previous example to get the $(6, 8, 10)$ triple. This, however, gives a similar triangle, one of exactly the same proportions, as the change is only a matter of scale and not of shape. Given the first triangle, we find the second Pythagorean triple simply by measuring the lengths of the sides in units that are half the size of the original units, thereby doubling the numerical size of the dimensions. There are, however, genuinely different triples such as those representing the $(5, 12, 13)$ and the $(65, 72, 97)$ right-angled triangles.

In order to describe all Pythagorean triples, therefore, it is enough to do the job for all triples (a, b, c) where the hcf of the three numbers is 1, as all others are merely scaled-up versions of these. The recipe is as follows. Take any pair of coprime positive integers m and n, with one of them even, and let m denote the larger. Form the triple given by $a = 2mn$, $b = m^2 - n^2$ and $c = m^2 + n^2$. The three numbers a, b, and c then give you a Pythagorean triple (the algebra is easily checked) and the three numbers have no common factor (also not difficult to verify). The three examples above arise by taking $m = 2$ and $n = 1$ in the first case, $m = 3$ and $n = 2$ in the second, while for the last triangle we have $m = 9$, $n = 4$. It takes more work to verify the converse: any such Pythagorean triple arises in this fashion for suitably chosen values of m and n, and what is more, the representation is unique so that two different pairs (m, n) cannot yield the same triple (a, b, c).

The corresponding equation for cubes and higher powers has no solution at all: for any power $n \geq 3$, there are no positive integer triples x, y, and z such that $x^n + y^n = z^n$. This is the famous *Fermat's Last Theorem*, which in future might be known as Wiles's

Theorem as it was finally proved in the 1990s by Sir Andrew Wiles. Even for the case of cubes, first solved by Euler, this is a very difficult problem. It is, however, relatively easy to show that the sum of two fourth powers is never a square (and so certainly not a fourth power). This is enough to reduce the problem to the case where n is a prime p (meaning that if we solved the problem for all prime exponents, the general result would follow at once), and indeed the problem was solved for so-called *regular* primes in the 19th century. However, the full solution was only realized as a consequence of Wiles settling a deep question called the Shimura–Taniyama Conjecture.

The most intensively investigated Diophantine equation is, however, the *Pell equation*, $x^2 - ny^2 = 1$, where n is a positive integer that is not itself a square. Its significance was appreciated very early, for it seems it was studied both in Greece and in India perhaps as far back as 400 BC, because its solution allows good rational approximations, $\frac{x}{y}$ of $\sqrt{n}$. For example, when $n = 2$, the equation has as one solution pair the numbers $x = 577$ and $y = 408$, and $(\frac{x}{y})^2 = 2.000006$. The equation is related to the geometric process of what the Greeks called *anthyphairesis* where one begins with two line segments and continues to subtract the shorter from the longer, a kind of analogue of the Euclidean Algorithm but applied to continuous lengths. Indeed, the Archimedes Cattle Problem mentioned in Chapter 5 leads to an instance of the Pell equation.

Versions of the Pell equation were studied by Diophantus himself around AD 150 but the equation was solved by the great Indian mathematician Brahmagupta (AD 628) and his methods were improved upon by Bhaskara II (AD 1150), who showed how to create new solutions from a seed solution. In Europe, it was Fermat who exhorted mathematicians to turn their attention to Pell's equation and the complete theory is credited to the renowned French mathematician Joseph-Louis Lagrange (1736–1813) (the English appellation 'Pell' is an historic accident).

The general method of solution is based on the continued fraction expansion of $\sqrt{n}$.

Fibonacci and continued fractions

Recall the sequence of numbers, $1, 1, 2, 3, 5, 8, 13, 21, \cdots$ discovered by Fibonacci and introduced in Chapter 5. Take a pair of successive terms in this sequence and write the corresponding ratio as 1 plus a fraction. If we now 'Egyptianize' this fraction by repeatedly dividing top and bottom by the numerator, a striking pattern emerges. Take, for instance

$$\frac{13}{8} = 1 + \frac{5}{8} = 1 + \frac{1}{1 + \frac{3}{5}} = 1 + \frac{1}{1 + \frac{1}{1 + \frac{2}{3}}} = 1 + \frac{1}{1 + \frac{1}{1 + \frac{1}{1 + \frac{1}{1 + \frac{1}{1}}}}}$$

We obtain a multi-floored fraction consisting entirely of 1s, and each preceding ratio of Fibonacci numbers appears as we wind through the calculation. This must happen every time: by the very way these numbers are defined, each Fibonacci number is less than twice the next, and so the result of the division will leave a quotient of 1 and the remainder is the preceding Fibonacci number. You will recall that the ratio of successive Fibonacci numbers approaches the Golden Ratio, τ, and so this suggests that τ is the limiting value of the continued fraction consisting entirely of 1s.

As was shown in Chapter 5, the value of an infinite repeating process may be made the subject of an equation based on that process. If we call the value of the infinite fractional tower of 1s by the name a, we see that a satisfies the relation $a = 1 + \frac{1}{a}$, because what lies underneath the first floor of the fraction is just another copy of a. From this, we see that a satisfies the quadratic equation $a^2 = a + 1$, the positive root of which is $\tau = \frac{1+\sqrt{5}}{2}$.

The type of continued fractions that emerge from this process are intrinsically important. When we approximate an irrational number y by rationals we naturally turn to the decimal representation of y. This is excellent for general calculations but, being tied to a particular base, is not mathematically natural. Essential to the nature of y is how well our number y can be approximated by fractions with relatively small denominators. Is there any way to find a series of fractions that best deals with the conflicting demands of approximating y to a high degree of accuracy while keeping the denominators relatively small? The answer lies in the continued fraction representation of a number that does this through its truncations at ever lower floors.

Continued fractions look very awkward because of the many floors we have used in representing them. However, the inconvenience of writing all the floors of the division is easily side-stepped – since every numerator is 1, we need only record the quotients in the division to specify which continued fraction we are talking about. For instance, the representation for the fraction $\frac{25}{91}$ develops as follows:

$$\frac{25}{91} = 0 + \cfrac{1}{3 + \frac{16}{25}} = 0 + \cfrac{1}{3 + \cfrac{1}{1 + \frac{9}{16}}} = \ldots$$

that eventually yields a continued fraction specified by the list [0, 3, 1, 1, 1, 3, 2]. As we have seen, the Golden Ratio, τ has the continued fraction representation [1, 1, 1, 1, $\cdots$]. In a fashion reminiscent of repeating decimal notation, we write $\tau = [\overline{1}]$. The first instance of 3 in the continued fraction of $\frac{25}{91}$ comes from writing $91 = 3 \times 25 + 16$, which is the first line in the Euclidean Algorithm for the pair (91, 25). Indeed, for this very reason there is one line in the continued fraction for every line of the algorithm when performed on the two numbers. In particular, starting with a *reduced fraction* in which the two numbers are coprime, the same will apply to each of the fractions that arise in the course of the calculation of the corresponding continued fraction.

The special example afforded by the Golden Ratio opens the door to the idea that we may be able to represent other irrational numbers not by finite continued fractions (which are obviously just rational themselves) but by infinite ones. But how is the continued fraction of a number a produced? The reader will need to tolerate a little algebraic trickery in order to see this in action, but here is how it goes.

There are two steps in the calculation of a continued fraction for a number $a = [a_0, a_1, a_2, \ldots]$. The number a_0 is the integer part of a, denoted by $a_0 = \lfloor a \rfloor$. (For example, the integer part of $\pi = 3.1415927 \cdots$ is given by $\lfloor \pi \rfloor = 3$.) In general, $a_n = \lfloor r_n \rfloor$, the integer part of r_n, where the remainder term r_n is defined recursively by $r_0 = a$, $r_n = \frac{1}{r_{n-1} - a_{n-1}}$. Applying this for example to $a = \sqrt{2}$ and employing the algebraic device of rationalizing the denominator (with which some readers may be familiar) we obtain, since $\lfloor \sqrt{2} \rfloor = 1$:

$$a = r_0 = \sqrt{2} = 1 + (\sqrt{2} - 1) \text{ so that } a_0 = 1;$$

$$r_1 = \frac{1}{r_0 - a_0} = \frac{1}{\sqrt{2} - 1} = \frac{\sqrt{2} + 1}{(\sqrt{2} - 1)(\sqrt{2} + 1)} = \sqrt{2} + 1, a_1 = \lfloor r_1 \rfloor = 2;$$

Next we get

$$r_2 = \frac{1}{r_1 - a_1} = \frac{1}{(\sqrt{2} + 1) - 2} = \frac{1}{\sqrt{2} - 1} = \frac{\sqrt{2} + 1}{(\sqrt{2} - 1)(\sqrt{2} + 1)} = \sqrt{2} + 1,$$

so that $r_1 = r_2 = \cdots = \sqrt{2} + 1$, $a_1 = a_2 = \cdots = 2$, and so $\sqrt{2} = [1, \overline{2}]$.

Indeed, the numbers that have recurring representations as continued fractions are rational numbers (which are exactly the ones whose representations terminate) and those that arise from quadratic equations such as τ, which we saw above is one solution of the equation $x^2 = x + 1$, and $\sqrt{2} = [1, \overline{2}]$, which satisfies $x^2 = 2$. Some other examples showing the rather unpredictable nature of the recurrences are $\sqrt{3} = [1, \overline{1, 2}]$, $\sqrt{7} = [2, \overline{1, 1, 1, 4}]$, $\sqrt{17} = [4, \overline{8}]$ and $\sqrt{28} = [5, \overline{3, 2, 3, 10}]$. There is nevertheless one very particular and remarkable facet to the pattern of the expansion of the continued fraction of an irrational square root. The expansion begins with an integer r, and the recurrent block consists of a palindromic sequence (a sequence of numbers that reads the same in reverse) followed by $2r$. This can be seen in all the preceding examples: for instance for $\sqrt{28}$ we see that $r = 5$, the palindromic part of the expansion is 3, 2, 3, which is followed by $2r = 10$. For $\sqrt{2}$ and $\sqrt{17}$, the palindromic part is empty, but the pattern is still there, albeit in a simple form. It can be shown that the continued fraction representation of a number is unique – two different continued fractions have different values.

The importance of continued fractions in approximation of irrationals by rationals is due to the so called *convergents* of the fraction, which are the rational approximations of the original number that result from truncating the representation at some point and working out the corresponding rational number. These represent the best approximation possible to the number in question in the sense that any better approximation will have a larger denominator than that of the convergents. The convergents of the Golden Ratio are the Fibonacci ratios. Since every term in the continued fraction representation of τ is 1, the convergence of these ratios is retarded as much as it possibly could be. For that reason, there is no more difficult number than τ to approximate by rationals and the Fibonacci ratios are the best you can do.

If the denominator of a convergent of a continued fraction is q, then the approximation is always within $\frac{1}{\sqrt{5}q^2}$ of the true value of the number and the convergents of a continued fraction alternately underestimate and overestimate the value to which they approach. It is, however, the euclidean numbers such as τ and $\sqrt{2}$ that are the worst when it comes to rational approximation. Some particular transcendentals, whose nature may seem the farthest removed from the rational world, may yet be approximated very closely and have convergents that home in on the target number with great rapidity.

The connection with Pell's equation, $x^2 - ny^2 = 1$, mentioned at the close of the previous section, now comes about as the solution (x, y) to the equation with the minimum possible positive value of x exists and is to be found among the convergents of the continued fraction representation of $\sqrt{n}$. For example, when $n = 7$ the sequence of convergents of $\sqrt{7}$ begins with $2/1$, $3/1$, $5/2$, $8/3$, $\cdots$ and it is $x = 8$, $y = 3$ that provides this smallest so called *fundamental solution* of the Pell equation $x^2 - 7y^2 = 1$. The fundamental solution, however, sometimes does not turn up at all early in the expansion: for example, the smallest positive solution to $x^2 - 29y^2 = 1$ is $x = 9801$ and $y = 1820$. Once this fundamental solution (x, y) has been located, however, all other solutions arise by taking successive powers of the expression $(x + y\sqrt{n})^k$ $(k = 1, 2, 3, \cdots)$ and extracting the coefficients of the corresponding rational and irrational parts of the expanded expression. In this way, the full solution set of the Pell equation is realized through the continued fraction representation of $\sqrt{n}$.

Chapter 8
Numbers but not as we know them

Real and complex numbers

The construction of the complex numbers is much simpler and goes much more smoothly than the construction of the real numbers. The first stage in producing the reals is development of the rationals, at which point we have to explain what is meant by a fraction. A fraction, such as $\frac{2}{3}$ is just a pair of integers, which we represent in this familiar but peculiar manner. The idea of fractional parts is not difficult to understand, although the corresponding arithmetic takes real effort to master. Along the way your teachers explain in passing that such fractions as $\frac{2}{3}$, $\frac{4}{6}$, $\frac{6}{9}$ etc. are 'equal' – they are not the same number pairs but they do represent equal slices of pie. This is not hard to accept but it does draw our attention to the fact that a rational number is in reality an infinite set of equivalent fractions, each represented by a pair of integers. This sounds intimidating and we might prefer not to think too much about this, for the prospect of manipulating infinite collections of pairs of integers might leave us feeling uneasy. There is one saving grace in that any fraction has a unique reduced representation where the numerator and denominator are coprime, which can be got by cancelling any common factors in the fraction with which you originally began. Nonetheless, once you are familiar with the properties of fractions and the rules for using them, nothing should go wrong even though closer examination reveals that, as you do your sums, you are implicitly manipulating infinite collections of integer pairs.

However, things get worse at the next stage when we try to pin down what real numbers truly are. Let's begin with Pythagoras' problem – he found that there was no fraction equal to $\sqrt{2}$, so we can introduce a new symbol r, endow it with the property that $r^2 = 2$, and form a new 'field' of numbers from the rationals together with the new number r. This works in that the set of all numbers of the form $a + br$, where a and b are rational obey all the normal rules of algebra – we can even divide because the reciprocal of a number of this form retains the form, as can be seen through a little fancy algebraic footwork known as rationalizing the denominator.

The new numbers r and $-r$ furnish the two solutions of the equation $x^2 = 2$, but what about $x^2 = 3$? It seems that we need to adjoin yet another new number in order to solve this equation as it is easy to check that no number of the form $a + br$ will square to give 3. (A simple contradiction argument suffices here: assuming that $(a + br)^2 = 3$, allows you to deduce the false statement that at least one of $\sqrt{2}$ or $\sqrt{3}$ is rational after all.)

It is tempting to cut through all this fretting about particular equations and simply declare that we already know what the real numbers are – they are the collection of all possible decimal expansions, both positive and negative. These are very familiar, in practice we know how to use them, and so we feel on safe ground. At least until we ask some very basic questions. The main feature of numbers is that you can add, subtract, multiply, and divide. But, for example, how are you supposed to multiply two infinite non-recurring decimals? We depend on decimals being finite in length so that you 'start from the right-hand end', but there is no such thing with an infinite decimal expansion. It can be done, but it is complicated both in theory and in practice. A number system where you struggle to explain how to add and multiply does not seem satisfactory.

And there are other little pitfalls. When you multiply $\frac{1}{3}$ by 3, the answer is 1. When you multiply $0.333\cdots$ by 3, the answer is surely $0.999\cdots$. It is indeed the case that two different decimal expansions can represent the same number: $1.000\cdots = 0.999\cdots$. In fact, this happens with any terminating decimal, for example $0.375 = 0.374999\cdots$. Hence it can't be quite right to say that decimals and real numbers are one and the same, as we see that two different decimal expansions can equal the same number. Moreover, the numbers with non-unique decimal expansions will change if we work in another base, and that raises another complication. If we define real numbers by decimals, we are making the construction depend on an arbitrary choice (base ten). If we do the same construction in binary, will the set of 'real numbers' be the same? And what do we mean by 'the same' in any case?

You may find the foundational questions raised above interesting or you may grow impatient with all the introspection as we seem to be making trouble for ourselves when previously all was smooth sailing. There is a serious point, however. Mathematicians appreciate that, whenever new mathematical objects are introduced, it important to construct them from known mathematical objects, the way, for instance, fractions can be thought of as pairs of ordinary integers. In this way, we may carefully build up the rules that govern the new extended system and know where we stand. If we neglect foundations completely, it will come back to haunt us later. For example, the rapid development of calculus, which was born out of the study of motion, led to spectacular results, such as prediction of the movement of the planets. However, manipulation of infinite things as if they were finite sometimes provided amazing insights and at other times patent nonsense. By putting your mathematical systems on a firm foundation, we can learn how to tell the difference. In practice, mathematicians often indulge in 'formal' manipulations in order to see if some sparkling new theorem is in the offing. If the outcome is worthy of attention, the result can be proved rigorously by going back to basics and by invoking results that have been properly established earlier.

This is why Julius Dedekind (1831–1916) took the trouble of formally constructing the real number system based on his idea that is now referred to as *Dedekind cuts* of the real line. The first mathematician, however, to successfully deal with the dilemma caused by the existence of irrational numbers was Eudoxus of Cnidus (fl 380 BC) whose *Theory of Proportions* allowed Archimedes to use the so-called *Method of Exhaustion* to rigorously derive results on areas and volumes of curved shapes before the advent of calculus some 1,900 years later.

The final piece of the number jigsaw – the imaginary unit

The introduction of a new symbol i that squares to -1 is staggeringly successful as it resolves at a stroke not only the problem of providing a solution to one equation but allows solution of *all* polynomial equations and much more besides. We certainly have two square roots of any negative number, $-r$, as both the numbers $\pm i\sqrt{r}$ square to $-r$ by virtue of the property that $i^2 = -1$ and the assumption that, as with ordinary arithmetic, multiplication *commutes* meaning that $zw = wz$ for any numbers z and w. Indeed, if we continue on the basis that the system of complex numbers $a + bi$ should subsume that of the reals (which correspond to the case where $b = 0$) and that all the normal rules of algebra should continue to be respected, we meet with no difficulties and many pleasant surprises. The set of complex numbers, denoted by **C** is a 'field' which, among other things, guarantees that division is also possible. To see how it all works out, however, it is best to leave the monorail of the real line and look at life in two dimensions.

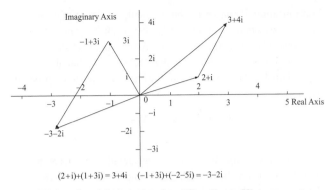

$(2+i)+(1+3i)=3+4i$ $(-1+3i)+(-2-5i)=-3-2i$

13. Addition of complex numbers by adding directed line segments

The arithmetic of complex numbers presents itself very nicely in the *complex plane*. We think of the complex number $a + bi$ as being represented by the point (a, b) in the coordinate plane. When we add two complex numbers $z = (a, b)$ and $w = (c, d)$, we simply add their first and second entries together, to give us $z + w = (a + c, b + d)$. If we make use of the symbol i, we have for example $(2 + i) + (1 + 3i) = 3 + 4i$.

This corresponds to what is known as *vector addition* in the plane, where directed line segments (*vectors*) are added together, top to tail (see Figure 13). We begin at the *origin*, which has coordinates of $(0, 0)$, and in this example we lay down our first arrow from there to the point $(2, 1)$. To add the number represented by $(1, 3)$, we go to the point $(2, 1)$, and draw an arrow that represents moving 1 unit right in the horizontal direction (that is the direction of the *real axis*), and 3 units up in the direction of the vertical (the *imaginary axis*). We end up at the point with coordinates $(3, 4)$. In much the same way, we can define subtraction of complex numbers by subtracting the real and imaginary parts so that, for example, $(11 + 7i) - (2 + 5i) = 9 + 2i$. This can be pictured as starting with the vector $(11, 7)$, and subtracting the vector $(2, 5)$, to finish at the point $(9, 2)$.

111

Multiplication is another matter. Formally it is easy to do: we multiply two complex numbers together by multiplying out the brackets, remembering that $i^2 = -1$. Assuming the *Distributive Law* continues to hold, which is the algebraic rule that allows us to expand the brackets in the usual way, then multiplication proceeds as follows:

$$(a + bi)(c + di) = a(c + di) + bi(c + di) =$$
$$ac + adi + bci + bdi^2 = (ac - bd) + (ad + bc)i$$

Division, on the other hand, can be calculated by means of the *complex conjugate*. In general, the conjugate of $z = a + bi$ is denoted by $\bar{z}$ and is $a - bi$, in other words, $\bar{z}$ is the reflection of z in the real axis. The multiplication rule applied to $z\bar{z}$ gives $a^2 + b^2$, which is just a real number as the imaginary part turns out to be zero. This equals the square of the distance of z from the origin, which is denoted by $|z|$. In symbols $z\bar{z} = |z|^2$. We may now divide one complex number by another by multiplying top and bottom by the conjugate of the divisor in order to make the division one by a purely real number. This is analagous to the standard technique of rationalizing the denominator that is used to remove square roots in the bottom line, which we used to calculate the continued fraction for $\sqrt{2}$. For example:

$$\frac{15 + 16i}{2 + 3i} = \frac{(15 + 16i)(2 - 3i)}{(2 + 3i)(2 - 3i)} = \frac{30 - 45i + 32i - 48i^2}{2^2 - 6i + 6i - 3^2 i^2}$$
$$= \frac{(30 + 48) - (45 - 32)i}{2^2 + 3^2} = \frac{78 - 13i}{13} = 6 - i.$$

By using general rather than specific complex numbers we can, in the same way, find the outcome of a general division of complex numbers in terms of their real and imaginary parts as we have done above for general complex multiplication. However, as long as the technique is understood, there is no pressing need to produce and to memorize the resulting formula.

Multiplication has a geometric interpretation that is revealed if we alter our coordinate system from the ordinary rectangular

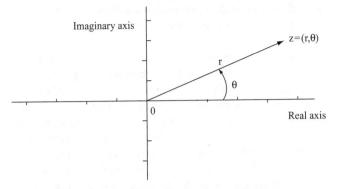

Imaginary axis

$z = (r, \theta)$

r

θ

0

Real axis

14. The position of a complex number in polar coordinates

coordinates to *polar coordinates*. In this system, a point z is once
again specified by an ordered pair of numbers, which we shall
write as (r, θ). The number r is the *distance* of our point z from the
origin O (called in this context the *pole*). Therefore r is a
non-negative quantity and all points with the same value of r form
a circle of radius r centred at the pole. We use the second
coordinate θ to specify z on this circle by taking θ to be the *angle*,
measured in an anti-clockwise direction, from the real axis to the
line Oz. The number r is called the *modulus* (plural moduli) of z,
while the angle θ is called the *argument* of z.

Suppose now that we have two complex numbers, z and w, whose
polar coordinates are (r_1, θ_1) and (r_2, θ_2) respectively. It turns out
that the polar coordinates of their product zw take on a simple and
pleasing form. The rule of combination can be expressed neatly in
ordinary language: the modulus of the product zw is the product
of the moduli of z and w, while the argument of zw is the *sum* of
the arguments of z and w. In symbols, zw has polar coordinates
$(r_1 r_2, \theta_1 + \theta_2)$. The multiplication of the real numbers is subsumed
under this more general way of looking at things: a positive real
number r, for instance, has polar coordinates $(r, 0)$, and if we

multiply by another $(s, 0)$, the result is the expected $(rs, 0)$, corresponding to the real number rs.

Much more of the character of the multiplication of complex numbers is revealed through this representation. The polar coordinates of the complex unit i are given by $(1, 90°)$. (Normally, angles are not measured in degrees in such circumstances but in the natural mathematical unit of the *radian*: there are 2π radians in a circle, so that a turn of one radian corresponds to moving one unit along the circumference of the unit circle, centred at the origin. One radian is about $57.3°$.) If we now take any complex number $z = (r, \theta)$ and multiply by $i = (1, 90°)$, we find that $zi = (r, \theta + 90°)$. That is to say, *multiplication by i corresponds to rotation through a right angle about the centre of the complex plane*. In other words, the right angle, that most fundamental geometric idea, can be represented as a number.

Indeed, the effect of adding or multiplying by a complex number z on all the points in a given region of the complex plane can be pictured geometrically. Imagine any region you fancy in the plane. If we *add* z to every point inside your region, we simply move each point the same distance and direction determined by the arrow, or vector as we often call it, represented by z. That is to say we *translate* the region to some other position in the plane so that the shape and size are exactly maintained, as is its attitude, by which we mean the region has not undergone any rotation or reflection. Multiplying every point in your region by $z = (r, \theta)$ has two effects, however, one caused by r and the other by θ. The modulus of each point in the region is increased by a factor r, so all the dimensions of the region are increased by a factor of r also (so its area is multiplied by a factor of r^2). Of course, if $r < 1$ then this 'expansion' is better described as a contraction as the new region will be smaller than the original. The region will, however, maintain its shape – for instance, a triangle is mapped on to a similar triangle with the same angles as before. The effect of θ, as we have explained above, is to rotate the region through an angle

114

θ, anticlockwise about the pole. The net effect then in multiplying all points of your region by z is to expand and rotate your region about the pole. The new region will still have the same shape as before but will be of a different size, determined by r, and will be lying in a different attitude as determined by the rotation angle θ.

Further consequences

The polar version of complex numbers is particularly suited to the taking of powers and roots for to raise $z = (r, \theta)$ to some positive power n, we simply raise the modulus to that power, and add θ to itself n times, to give $z^n = (r^n, n\theta)$. The same formula applies to fractional and negative powers. Division can be comprehended in polar form as well. As with real numbers, division by a complex number z means multiplication by its reciprocal $w = \frac{1}{z}$, but what number is this w? Given that $z = (r, \theta)$ the number w is the one with the property that $zw = (1, 0)$, the number 1. This shows us that we must take $w = (\frac{1}{r}, -\theta)$, for then $zw = (r, \theta)(\frac{1}{r}, -\theta) = (r\frac{1}{r}, \theta - \theta) = (1, 0)$, as we require. This gives an alternative to the rectangular approach to division that makes use of complex conjugation.

There are a host of applications of complex numbers, even at the elementary level. The interplay between rectangular and polar representations brings trigonometry into play in a surprising and advantageous way. For instance, a standard exercise for students is the derivation of important identities that now arise very naturally by taking arbitrary complex numbers of unit modulus (i.e. $r = 1$), and calculating powers using both rectangular and then polar coordinates. Equating the two forms of the answer then reveals a trigonometric equation.

A point with polar coordinates $(1, \theta)$ has, by elementary trigonometry, the rectangular coordinates $(\cos \theta, \sin \theta)$. If we now multiply two such complex numbers $z = \cos \theta + i \sin \theta$ and $w = \cos \phi + i \sin \phi$ in rectangular coordinates we obtain:

$$zw = (\cos\theta\cos\phi - \sin\theta\sin\phi) + i(\cos\theta\sin\phi + \sin\theta\cos\phi)$$

while the same in polar coordinates gives:

$$zw = (1, \theta)(1, \phi) = (1, \theta + \phi) = \cos(\theta + \phi) + i\sin(\theta + \phi);$$

equating the real and imaginary parts of the two versions of this one product then painlessly yields the standard angle sum formulas of trigonometry:

$$\cos(\theta+\phi) = \cos\theta\cos\phi - \sin\theta\sin\phi, \ \ \sin(\theta + \phi) = \cos\theta\sin\phi + \sin\theta\cos\phi.$$

Alternatively, the polar form for complex multiplication can be derived using these trigonometric formulas. Indeed, the rule that we have stated here, without proof, for multiplication in polar form is usually first derived from the rectangular form by using trigonometric formulas.

Much more comes quite easily now as the use of complex numbers reveals a connection between the exponential or power function, and the seemingly unrelated trigonometric functions. Without passing through the portal offered by the square root of minus one, the connection may be glimpsed, but not understood. The so-called *hyperbolic functions* arise from taking what are known as the even and odd parts of the exponential function. To every trigonometric identity there corresponds one of identical form, except perhaps for sign, involving these hyperbolic functions. This can be verified easily in any particular case, but then the question remains as to why it should happen at all. Why should the behaviour of one class of functions be so closely mirrored in another class, defined in so different a manner, and of such different character? Resolution of the mystery is by way of the formula $e^{i\theta} = \cos\theta + i\sin\theta$, which shows that the exponential and trigonometric functions are intimately linked, but only through use of the imaginary unit i. Once this is revealed (for it is surprising and is by no means obvious), it becomes clear that results along the lines described are inevitable through performing calculations using the two alternative representations offered by this equation and then equating real and imaginary parts. Without the formula, however, it all remains a mystery.

Complex numbers and matrices

Let us examine some consequences of the revelation that multiplication by i represents a rotation through a right angle about the centre of the coordinate plane. If $z = x + iy$, we have through expanding the brackets and reordering multiplications that $i(x + iy) = -y + ix$, so that the point (x, y) is taken to $(-y, x)$ under this rotation; see Figure 15. In this way, multiplication by i can be regarded as operating on points in the plane. This operation enjoys the special property that for any two points z and w and any real number a, we have $i(z + w) = iz + iw$, and $i(aw) = a(iw)$.

Moreover, if we multiply a real number a by a complex number $x + iy$, we get $a(x + iy) = ax + i(ay)$. In terms of points in the

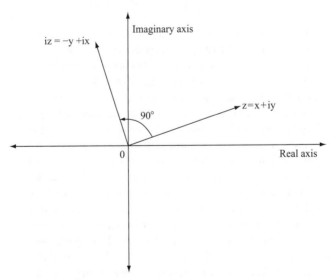

15. Multiplication by i rotates a complex number by a right angle

complex plane, we have that (x, y) is moved to (ax, ay), or to write it another way, $a(x, y) = (ax, ay)$.

The kinds of operations that enjoy these two properties are known as *linear* and are of paramount importance throughout all mathematics. Here, I wish only to draw to your attention to the fact that the effect of such an operation L is determined by its action on the two points $(1, 0)$ and $(0, 1)$, for let us suppose that $L(1, 0) = (a, b)$ and $L(0, 1) = (c, d)$. Then for any point (x, y) we have $(x, y) = x(1, 0) + y(0, 1)$, and so using the properties of a linear operation we obtain:

$$L(x, y) = L(x(1, 0) + y(0, 1)) = xL(1, 0) + yL(0, 1) =$$
$$= x(a, b) + y(c, d) = (ax, bx) + (cy, dy) = (ax + cy, bx + dy).$$

This information may be summarized by what is known as a *matrix equation*:

$$(x, y) \begin{pmatrix} a & b \\ c & d \end{pmatrix} = (ax + cy, bx + dy).$$

Here we have drawn out an example of matrix multiplication, which indicates how that operation is carried out in general. A *matrix* is just a rectangular array of rows and columns of numbers. Matrices, however, represent another kind of two-dimensional numerical object and, what is more, they pervade nearly all of higher mathematics, both pure and applied. They represent a whole corpus of algebra, and much of modern mathematics strives to represent itself through matrices, so useful have they proved to be. Two matrices with the same number of rows and the same number of columns as each other are added entry-to-entry: for example, to find the entry in the second row and third column of the sum of two matrices, we simply add the correspondingly placed entries in the two matrices in question. It is matrix multiplication, however, that gives the subject a new and important character, and how it is conducted has emerged of its own accord in the previous example – each entry in the product matrix is formed by taking the *dot product* of a row of the first matrix with a column of the second, meaning that the entry is the sum of the corresponding products when the row of the first matrix is placed on top of the column of the second.

Matrices follow all the usual laws of algebra except commutativity of multiplication, meaning that for two matrices A and B it is *not* generally true that $AB = BA$. However, matrix multiplication is *associative*, meaning that products of any length may be written unambiguously without the need for bracketing.

Linear transformations of the plane are typically rotations about the origin, reflections in lines through the origin, enlargments and contractions about the origin, and so called *shears* (or *slanting*), which move points parallel to a fixed axis by an amount proportional to their distance from that axis in a manner similar to the way the pages of a book can slide past one another. Any sequence of these transformations can be effected by multiplying all of the relevant matrices together to reveal a single matrix that has the same net effect as all those transformations acting in turn. The rows of the resultant matrix are simply the images of the two points (1, 0) and (0, 1), as we saw above, known as *basis vectors*.

It is now natural to look at the matrix J that represents an anticlockwise rotation of a right angle about the origin as it should mimic the behaviour we see when we multiply by the imaginary unit i. Since the point (1, 0) is taken onto the point (0, 1) by the rotation and similarly the point (1, 0) moves to $(-1, 0)$, these two vectors form the rows of our matrix J. The result of squaring J will be a matrix that has the geometric effect of rotating points through $2 \times 90° = 180°$ about the origin. We calculate this below by matrix multiplication. To find, for example, the bottom right entry of J^2 we take the dot product of the second row and second column, which gives $(-1) \times 1 + 0 \times 0 = -1 + 0 = -1$. The complete calculation has the following outcome:

$$J^2 = \begin{pmatrix} 0 & 1 \\ -1 & 0 \end{pmatrix} \begin{pmatrix} 0 & 1 \\ -1 & 0 \end{pmatrix} = \begin{pmatrix} -1 & 0 \\ 0 & -1 \end{pmatrix} = -I.$$

The matrix I with rows $(1\,0)$ and $(0\,1)$ is the *identity matrix*, so called as it acts like the number 1 in that when multiplied by another matrix A the result is A. The matrix $-I$, which represents a full half turn rotation about the origin, does behave like -1 in that $(-I)^2 = I$. The upshot of all this is that the matrices $aI + bJ$, where a and b are real numbers, faithfully mimic the complex numbers $a + bi$ with respect to addition and multiplication, and so give a matrix representation of the complex number field. The matrix corresponding to the typical complex number $a + bi$ is

$$\begin{pmatrix} a & b \\ -b & a \end{pmatrix}.$$

The matrices that represent the complex numbers do commute with one another but, as was mentioned above, this does not generally apply to all matrix products and another way in which matrices can misbehave is that not all of them can be 'inverted'. For most square matrices A (a matrix with equal numbers of rows and columns), we may find a unique *inverse matrix B* such that $AB = BA = I$, the identity matrix. The existence of the inverse matrix however depends upon a single number associated with a square matrix known as its *determinant*. In general, this is a certain sum of signed products formed by taking one entry from each row and column of the array. For the typical 2×2 matrix array as introduced on page 118, the determinant is the number $\Delta = ad - bc$. Determinants have many uses and agreeable properties. For instance, Δ represents the area scale factor of the corresponding matrix transformation: a shape of area a will be transformed into one of area Δa when undergoing a transformation by that matrix (and if Δ is negative, the shape also undergoes a reflection, reversing the original orientation). What is more, the determinant of the product of two square matrices is the product of the determinants of those matrices. A square matrix A will have an inverse B except in the case where $\Delta = 0$, in which case it will not. A zero determinant corresponds geometrically to a *degenerate* transformation where areas are collapsed by the matrix to figures of zero area such as a line segment or even a single point.

For the matrix of a complex number $z = a + bi$, we note that $\Delta = a^2 + b^2$, which is never zero except when $z = 0$ – but of course the number 0 never had a reciprocal before, and that remains the case in the wider arena of the complex numbers. This does confirm however that every *non-zero* complex number possesses a multiplicative inverse.

We stand here on the edge of the vast worlds of linear algebra, representation theory, and applications to multi-dimensional calculus, and this is not the place to go further. However, the reader should be aware that matrices apply to three dimensions and indeed to n-dimensional space, typically through $n \times n$ matrices. Although the arrays become larger and more complicated, the matrices themselves yet remain two-dimensional numerical objects.

Numbers beyond the complex plane

The field **C** of all complex numbers is *complete* in two important ways. An infinite sequence of complex numbers in which the terms cluster into ever smaller circles of radius that approaches 0 is called *convergent*. Any convergent sequence of complex numbers approaches a limiting complex number. This is also true of the real numbers, but not of the rationals – the successive decimal approximations to any irrational number represent a sequence of rational numbers that approach a limit outside of the rationals. Moreover, **C** is complete (or *closed*) in the algebraic sense that it can be shown that any polynomial equation $p(z) = a + bz + cz^2 + \cdots + z^n = 0$ has n (complex) solutions, $z_1, z_2, \cdots, z_n$, which then allows $p(z)$ itself to be fully factorized as $p(z) = (z - z_1)(z - z_2) \cdots (z - z_n)$.

This and other stunning successes of the complex numbers largely obviate the need to expand the number system further beyond the complex plane. Indeed, it is not possible to construct an augmented number system that contains C and also retains all the normal laws of algebra. Moreover, there are only two extended systems that retain much algebraic structure at all, these being the *quaternions* and the *octonions*. Although their use is not nearly so widespread as that of the complex numbers, the *quaternions* are put to work, for example, in three-dimensional computer graphics. The octonions, which can be thought of as pairs of quaternions, lack not only the commutative property but also the associative property of multiplication.

A quaternion is a number of the form $z = a + bi + cj + dk$, where the first part $a + bi$ is an ordinary complex number and the two *quaternion units* j and k also satisfy $j^2 = k^2 = -1$. In order to do multiplication with quaternions, we need to know how the units multiply with one another and this is determined by the rules $ij = k$, $jk = i$, $ki = j$ but the reversed products carry the opposite sign, so that, for example, $ji = -k$ (indeed, all these products may be derived from the single additional equation: $ijk = -1$). The quaternions then form an enhanced algebraic system that satisfies all the laws of algebra except for commutativity of multiplication, due to the sign changes mentioned above in the reversed products. The consistency of the system can also be demonstrated through representation by 2×2 matrices, but this time we allow complex rather than just real entries. The number 1 is once more identified with I, the identity matrix but the units i, j, and k have as their matrix counterparts:

$$i = \begin{pmatrix} i & 0 \\ 0 & -i \end{pmatrix} \quad j = \begin{pmatrix} 0 & 1 \\ -1 & 0 \end{pmatrix} \quad k = \begin{pmatrix} 0 & i \\ i & 0 \end{pmatrix}$$

while the typical quaternion z has as its matrix:

$$z = \begin{pmatrix} a + bi & c + di \\ -c + di & a - bi \end{pmatrix}.$$

This representation of the quaternions by matrices is not unique, however, and indeed the representation of the complex numbers by matrices also has equivalent alternatives. Moreover, it is possible to represent the quaternions without employing complex numbers but only at the expense of using larger matrix arrays: the quaternions can be represented by certain 4×4 matrices with only real number entries.

New kinds of numbers and the extensions of old systems have come about through the need to perform calculations the outcome of which could not be accommodated by the number system as it stood. Every civilization begins with the counting numbers, but calculations involving fragments lead to fractions, those involving debt lead to negatives, and as Pythagoras discovered, those involving lengths lead to irrational numbers. Although a very ancient revelation, the fact that not all numerical matters could be dealt with using whole numbers and their ratios was a subtle discovery of a deeper kind. As science became more sophisticated, the number systems required have needed to mature in order to deal with these advances. Scientists do not generally look to create new numbers systems in a whimsical fashion. On the contrary, they are introduced often reluctantly and hesitatingly at first, to deal with research problems. For example, although first introduced in the 19th century, matrices arose irresistibly in quantum mechanics in the early 20th century when physicists encountered a quantity of the form $q = AB - BA$ that was nevertheless not zero. In any commutative system of numbers, q would of course be 0, so the numerical objects needed here were not of a kind they had met before: they were matrices.

It seems now that the world of mathematics and physics has enough number types. Although there are kinds of numbers not mentioned in this book, the number types that are commonly used throughout mathematics and science have not needed to change a great deal since the first half of the 20th century.

These observations, however, bring our mathematical balloon ride to its conclusion. We began at ground level and have ascended to where I hope the reader can gaze down upon a view of the rich and mysterious world of numbers.

Further reading

Two other books in the OUP VSI series that complement and expand on the current one are *Mathematics* by the Field's medallist Timothy Gowers and *Cryptography* by Fred Piper and Sean Murphy. Probability and statistics, fields that were neglected here in *Numbers*, are the subject of the *VSI Statistics* by David J. Hand.

An insight into the nature of numbers can be read in David Flannery's book, *The Square Root of 2: A Dialogue Concerning a Number and a Sequence* (Copernicus Books, 2006). This leisurely account is in the Socratic mode of a conversation between a teacher and pupil. *One to Nine: The Inner Life of Numbers* by Andrew Hodges (Short Books, 2007) analyses the significance of the first nine digits in order. Actually it uses each number as an umbrella for examining certain fundamental aspects of the world and introduces the reader to all manner of deep ideas. This contrasts with Tony Crilly's *50 Mathematical Ideas You Really Need To Know* (Quercus Publishing, 2007), which does as it says, digesting each of 50 notions into a four-page description in as straightforward a manner as possible. The explanations are mainly through example with a modest amount of algebraic manipulations involved, rounded off with historical details and timelines surrounding the commentary. A particularly nice account on matters concerned with binomial coefficients is the paperback of Martin Griffiths, *The Backbone of Pascal's Triangle* (UK Mathematics Trust, 2007), in which you will read proofs of Bertrand's Postulate and Chebyshev's Theorem, giving bounds for the number of primes less than n.

Elementary Number Theory by G. and J. Jones (Springer-Verlag, 1998) gives a gentle but rigorous introduction and goes as far as aspects of the famous Riemann Zeta Function and Fermat's Last Theorem. The classic book *An Introduction to the Theory of Numbers*, by G. H. Hardy and E. M. Wright, 6th edn (Oxford University Press, 2008) assumes little particular mathematical knowledge but hits the ground running. The author's book *Number Story: From Counting to Cryptography* (Copernicus Books, 2008) has more in the way of the history of numbers than this *VSI* and includes mathematical details in the final chapter. *The Book of Numbers* by John Conway and Richard Guy (Springer-Verlag, 1996) is full of history, vivid pictures, and all manner of facts about numbers. Quite a lot of the history and mystery surrounding complex numbers is to be found in *An Imaginary Tale: The Story of $\sqrt{-1}$* (Princeton University Press, 1998) by Paul J. Nahin. Paul Halmos's *Naive Set Theory* (Springer-Verlag, 1974) gives a quick mathematical introduction to infinite cardinal and ordinal numbers, which were not introduced here.

A popular account of the Riemann Zeta Function is the book by Marcus du Sautoy, *The Music of the Primes, Why an Unsolved Problem in Mathematics Matters* (HarperCollins, 2004), while Carl Sabbagh's, *Dr Riemann's Zeros* (Atlantic Books, 2003) treats essentially the same topic.

There are two accounts of the solution to Fermat's Last Theorem, those being *Fermat's Last Theorem: Unlocking the Secret of an Ancient Mathematical Problem* by Amir D. Aczel (Penguin, 1996) and *Fermat's Last Theorem* by Simon Singh (Fourth Estate, 1999). The best popular book on the history of coding up to the RSA cipher is also an effort of Simon Singh: *The Code Book* (Fourth Estate, 2000). The unsolvability of the quintic (fifth-degree polynomial equations) was not explained in our text here but is the subject of an historical account: *Abel's Proof: An Essay on the Sources and Meaning of Mathematical Unsolvability* (MIT Press, 2003) by Peter Pesic.

Websites

A very high-quality web page that allows you to dip into any mathematical topic, and is especially rich in number matters, is Eric Wolfram's *MathWorld: mathworld.wolfram.com*. For mathematical history topics, try *The MacTutor History of Mathematics archive* at St Andrews University, Scotland: *www-history.mcs.st-andrews.ac.uk/history.index.html*. Web pages accessed 8 October 2010. Wikipedia's treatment of mathematics by topic is generally serious and of good quality, although the degree of difficulty of the treatments is a little variable. For example, Wikipedia gives a good quick overview of important topics such as matrices and linear algebra.

Index

Expand your collection of
VERY SHORT INTRODUCTIONS

READING GUIDES

Very Short Introductions

Whether you are part of a reading group wanting to discuss non-fiction books or you are eager to further your thinking on a *Very Short Introduction*, these reading guides, written by our expert authors, will provoke discussions and help you to question again, why you think what you think.

STATISTICS
A Very Short Introduction
David J. Hand

Modern statistics is very different from the dry and dusty discipline of the popular imagination. In its place is an exciting subject which uses deep theory and powerful software tools to shed light and enable understanding. And it sheds this light on all aspects of our lives, enabling astronomers to explore the origins of the universe, archaeologists to investigate ancient civilisations, governments to understand how to benefit and improve society, and businesses to learn how best to provide goods and services. Aimed at readers with no prior mathematical knowledge, this *Very Short Introduction* explores and explains how statistics work, and how we can decipher them.

www.oup.com/vsi